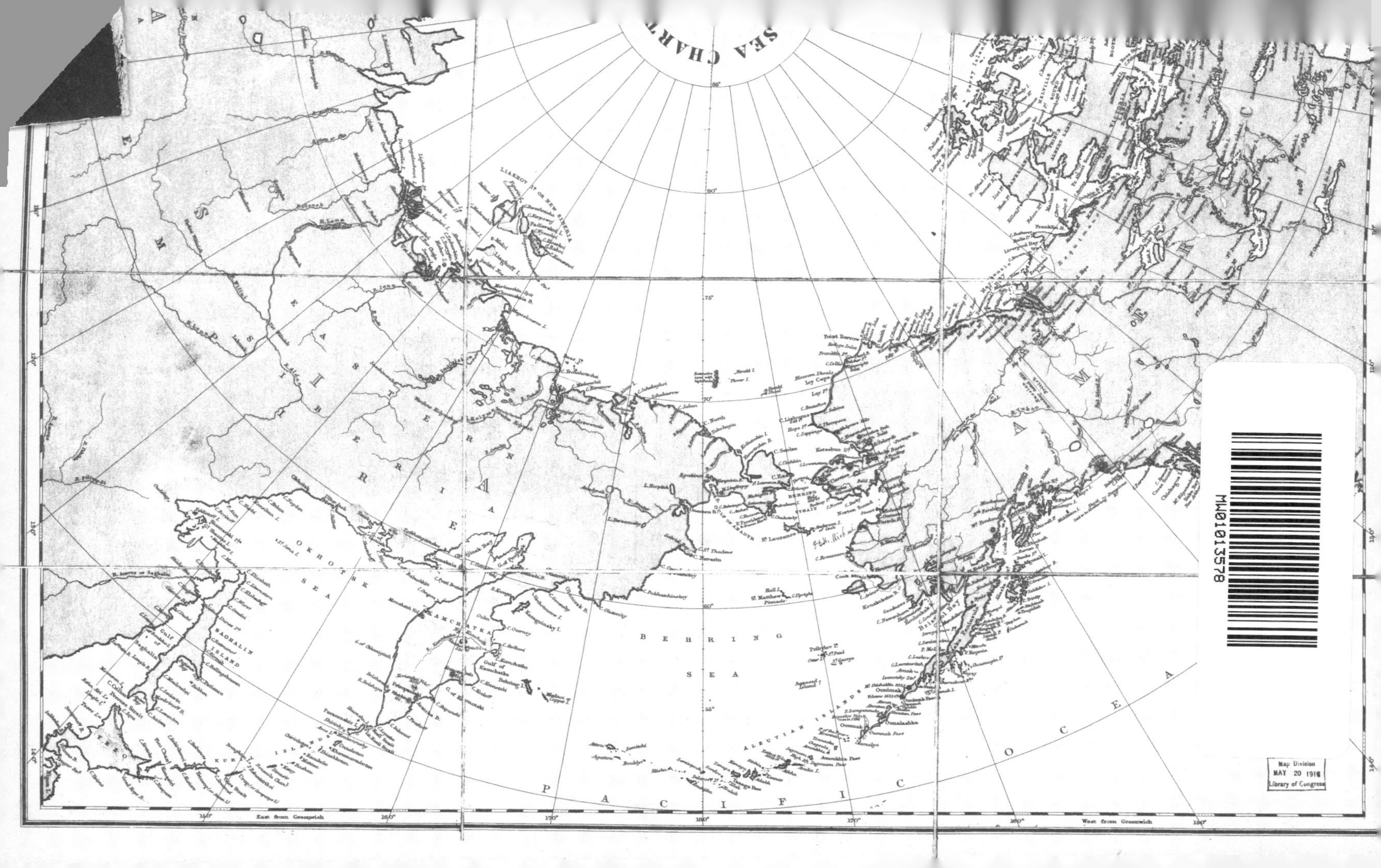

SEA CHART
LIAKHOV IS. OR NEW SIBERIA
EASTERN SIBERIA
OKHOTSK SEA
KAMCHATKA
SAGHALIN ISLAND
BEHRING SEA
ALEUTIAN ISLANDS
PACIFIC OCEAN
BEHRING STRAIT
St. Laurence
Point Barrow
Icy Cape
Kotzebue Sd.
Norton Sound
Bristol Bay
Cook Strait
Oumalashka
Oumnak
St. Matthew
Gulf of Kamchatka
Gulf of Saghalin
R. Lena
MELVILLE SOUND
VICTORIA
140° East from Greenwich 150° 160° 170° 180° 170° 160° West from Greenwich 150°

MORE PRAISE FOR *AFTER THE NORTH POLE*

"Exhilaratingly Epic as the Land in Grandeur, called HOME."

—dg nanouk okpik, American Book Award–winning author of *Corpse Whale*

"Layered into Erling Kagge's rich, evocative history of our quest to reach the North Pole—a quest characterized by dreams of conquest and exploitation, and ideas of romantic heroism, and tales of pain and suffering—is Kagge's own quest for the North Pole. We share his journey of beauty and fragility amid biting cold, gnawing hunger, and a life-threatening encounter with a polar bear. His story, of heaven and hell in one turn after another, is riveting."

—Deborah Cramer, author of *The Narrow Edge*

"A beautiful book which is as much about our inner world as our outer one, a book that helps us find the vast calm empty spaces within us and returns us to a feeling of being at home in the world."

—Alain de Botton, author of *The Consolations of Philosophy*

"It's impossible to read Erling Kagge and not fall in love with the North Pole. His enthusiasm is infectious, and his adventures are epic."

—Elizabeth Kolbert, bestselling author of *The Sixth Extinction*

"Erling Kagge charts a wondrous way to the North Pole, exploring the marvel and complexity of a place that exists both in reality and in our symbolic imagination. A fascinating and richly detailed book."

—Paul Lynch, Booker Prize–winning author of *Prophet Song*

"Kagge honors all those who survived or perished in the Arctic Circle, many he rescued from obscurity, with incredible tales of heroism, romance, and tragedy. . . . I found myself fascinated by the histories of Kagge's fellow explorers, who, like him, were blessed or cursed with a calling to trek there."

—Griffin Dunne, author of *The Friday Afternoon Club*

"I've lived the harshness of the polar regions, and Kagge captures it perfectly. *After the North Pole* brings you face-to-face with the raw, untamed beauty and unforgiving nature of the Arctic."

—Mike Horn, adventurer, explorer, and author of *Conquering the Impossible* and *Latitude Zero*

"Erling Kagge is a deeply thoughtful writer who has a strong and very individual style. The North Pole proves to be the perfect subject for him."

—Sir Michael Palin, English actor and author of *Erebus* and *Pole to Pole*

"With Erling Kagge alive and active, I have the satisfaction of knowing that I am not the only madman alive at present. He is also, never mind his icy achievements, a brilliant author."

—Sir Ranulph Fiennes, renowned British explorer and author of *Shackleton*

"Erling Kagge is an adventurer in the true sense of the word, exploring his curiosity with physical and intellectual and creative capabilities, all of which are prodigious, and fully engaging the reader in his adventures. At the conclusion of a Kagge read, I usually note my hands are sweating."

— Bob Shaye, founder of New Line Cinema

AFTER THE NORTH POLE

ALSO BY ERLING KAGGE

Silence: In the Age of Noise

Walking: One Step at a Time

Philosophy for Polar Explorers

HARPERONE

An Imprint of HarperCollins*Publishers*

AFTER THE NORTH POLE

A Story of Survival, Mythmaking, and Melting Ice

Erling Kagge

Translated by Kari Dickson

Page iv–v: A photo taken by Anthony Fiala, part of the Ziegler expedition, in March 1905 at 82° N. Lat.

Page vii: Men with dogsleds on an expanse of frozen ice during the Cook expedition to the North Pole, around 1908.

Page viii: A long-lost love of explorer Nils Strindberg: Anna Charlier, died 1949.

The credits on pages 345–46 constitute a continuation of this copyright page.

HarperCollins books may be purchased for educational, business, or sales promotional use. For information, please email the Special Markets Department at SPsales@harpercollins.com.

Originally published as *NORDPOLEN Natur, myter, eventyrlyst og smeltende is* in Norway in 2024 by Kagge Forlag

FIRST HARPERONE HARDCOVER PUBLISHED 2025

A similar edition of this book will be published by Penguin Random House in the UK as *North Pole: The History of an Obsession*

Designed by Bonni Leon-Berman

Library of Congress Cataloging-in-Publication Data has been applied for.

ISBN 978-0-06-342178-3

24 25 26 27 28 LBC 5 4 3 2 1

To my three daughters,
Ingrid, Solveig, and Nor,
and to Geir Randby

Most people are satisfied too soon, and that is the reason there is so little wisdom in the world.

FRIDTJOF NANSEN, POLAR EXPLORER

CONTENTS

Erling straddling a lead on his way to the North Pole.

INTRODUCTION

WHEN MY FRIEND BØRGE OUSLAND and I reached the North Pole on May 4, 1990, we were the first people to reach it on skis, without dogs, depots, and motorized aids. For fifty-eight days we had skied alone; sometimes we talked, but more often we silently got up early each day and put one foot in front of the other.

For years, I was convinced that the most important thing about the North Pole was to get there on skis. To win a race, to set a record. Our preparations took over two years, and during that time I thought only of one thing: to reach the North Pole. It might sound strange, but it was like a love affair—I thought of it as soon as I woke in the morning, throughout the day, and when I went to bed at night. I read and memorized, I skied whenever there was snow. And when there was no snow, I roller-skied, dragging tractor tires behind me.

"I'LL SEE YOU back here in two days," the pilot said matter-of-factly, as we landed by the ice edge on the north side of Ellesmere Island, Canada's northernmost island, on March 8 the same year. The airplane thermometer read –62°F. After fifteen years of flying in the Canadian Arctic, the pilot was used to seeing expeditions sputter out and come to nothing. It was early afternoon, and the sun was already setting to the south-southwest as we dragged our pulks[1] from the plane. In the low sunlight, we looked north, into the darkness, toward the pack ice. Some 478 miles ahead of us, at the North Pole, the sun had not risen for a full six months.

The sun emitted no warmth. In early March, temperatures remain more or less the same, whether the sun shines or not. Even when the sky is petrol-blue, the sun is so low and the light so weak that you barely cast a shadow. It is so cold that any particle carrying smell, bacteria, pollution, precipitation, or moisture quickly turns to ice. But to my surprise, I suddenly caught the scent of flowers. For a moment, I was baffled, before I realized that Patricia, the copilot, had sprayed on some perfume. The smell took off with the plane, leaving nothing but the cold.

The pulks carried everything we thought we'd need. Sixty days' worth of dried meat, oats, fat, chocolate, and formula milk, as well as enough camping gas to fuel our Primus stove for seventy days. We didn't know how many days the expedition might last, and in case we ran out of food, the plan was to have enough fuel to melt ice and survive the last stretch by drinking water alone. Each pulk weighed 265 pounds. In addition to our own sleeping bags, we had a double sleeping bag so we could sleep together to conserve the heat, camping mats, the stove, maps, a tent, 2.7 pounds of tools, and, in case of a polar bear attack, two Smith & Wesson .44 Magnum revolvers.

As we would find out, nothing had been packed in vain.

It was only when we finally reached the North Pole, fifty-eight days after smelling Patricia's floral perfume, that I realized I had been wrong about the North Pole. My years of obsession were not with the Pole itself, but with overcoming danger, pain, cold, and hunger.

THE NORTH POLE is the navel of the world. All humanity, the oceans, landmasses, and continents turn almost imperceptibly around this single fixed point. When you look up at the night sky in the northern hemisphere, at the sea of stars, you can see that not only our planet, but all the stars move in a circle around the top of the world. With one exception: the North Star.

Due to its position, the history of the North Pole is different from the history of every other place on earth. The more I see, read, and learn about this place, the clearer it seems that the history of the North Pole is the story of our ongoing relationship with nature—our changing feelings and respect for an environment that has not been created by or for people.

The history of the North Pole has been carved and chiseled by the blind hand of nature. But it is also about people's dreams, projections, and their desire to control and exploit the natural world.

For more than 2.7 million years, the North Pole has been covered by ice, floating atop an ocean that is as deep as 15,771 feet in some places. But the ice is showing increasing vulnerability. It is the place that is heating up most rapidly on earth.

People's fascination, dreams of fame and fortune, and nations' competition for prestige and strategic advantages have been—and still are—the biggest draws to the top of the world. As are greed and trickery. These forces have intensified over recent decades—as the pack ice melts.

For me, the story of the North Pole has come to encapsulate two approaches to life. One is to let wonder be the very engine of life.

And the other is an approach best described by the Norwegian word *eventyrlyst*, a compound noun composed of two words: *eventyr*, which means "adventure," "story," or "fairy tale," and *lyst*, meaning "lust" or "desire." I have never heard a word in any other language that precisely combines imagination with thirst for new experiences.

HISTORIANS GENERALLY BASE their accounts of a time or event on what people experienced or saw. But the North Pole's history is unusual, as the first polar explorers did not travel north in person, but journeyed there in their imagination, and pondered the vast unknown that this place was for thousands of years.

In prehistoric times, through the Stone Age, the Bronze Age, the Iron Age, and the Middle Ages, people tried to gain an understanding of the North Pole by studying the stars, listening to each other's ideas, and freeing their imaginations. Astronomers, astrologists, geographers, and philosophers asked: What was the light like up there, the colors, the vegetation? Was there a kind of mainland in the most northern area of the world? Did people and animals live there? The answers were often surprisingly uniform. The North Pole was a shiny, magnetic mountain closest to the gods. The North Pole was a lost paradise, home to Adam and Eve. The North Pole had been, and perhaps still was, bright, warm, fertile.

In the following centuries, through the Renaissance and Enlightenment, there was a growing appetite in Europe to explore the unknown world. After understanding the rotation of the earth by studying the sky, the Prussian scholar Nicolaus Copernicus proposed a heliocentric system; he argued that the world was a planet like any other and demonstrated that the sun, not the earth, was the center of the solar system. The source of knowledge was human observation; it was no longer God-given. The Italian philosopher Giordano Bruno tried to convince his peers that there were countless solar systems, and that the universe was infinite. He was burned at the stake for heresy. In the sixteenth and seventeenth centuries, Portuguese and Spanish sailors crossed the Atlantic Ocean, the Pacific Ocean, and the Indian Ocean and came home with tales of unknown lands, people, oceans, and trading opportunities. Even as the world started to be mapped, what was hidden at the North Pole remained a mystery. And people continued to speculate.

For many men, wondering was not enough. Instead they followed their *eventyrlyst*. Until recently, polar explorers were exclusively men—Inuits, Europeans, and Americans—who tried to reach the North Pole on foot, on skis, by boat, balloon, and plane. They fought against ocean currents, the cold, storms, rain, wild animals, snow, and ice. Not only was the northern tip of the world one of the last confirmed places on earth to be reached in 1909, but these extreme expeditions also caused tremendous human suffering, and the ice, wind, water, and cold have claimed countless lives.

Throughout history, we have mostly been wrong about the North Pole. Until 1895, when Fridtjof Nansen and Hjalmar Johansen achieved a record Farthest North latitude of 86°13.6′ N, more was known about Venus, Mars, Jupiter, and the light side of the moon than the area around the North Pole. Men had studied the planets night after night for thousands of years, but no one had seen the North Pole.

THOSE WHO WANT to go to the North Pole do so for several reasons. I wanted to do something extreme (what exactly that might be was of secondary importance). The urge to break with daily routines at school and work and pit myself against nature, to challenge it, was overwhelming. I climbed mountains, walked, and skied greater and greater distances, sailed across the Atlantic, and eventually to Antarctica. Throughout these trips I was slowly falling in love with the dream of the North Pole. Like all polar explorers before me, I was also driven by a need for recognition, but also other reasons I still struggle to understand.

For as long as I can remember, daily life in the Kagge household of my childhood was based on "masculine values." The most revered feat that any member of the family could achieve was to cross-country ski as far as possible; to set a personal best for the day, the week, or the season. As my mother, Aase, once said, with resignation, in our house there are three normal seasons—spring, summer, and autumn—and then there is the season. More than anything else, I yearned for recognition from my father, Stein. I did not think about it at the time but have since felt great sympathy for my mother, who lived within the rigid, testosterone-driven value system of three sons and a husband, so different to her own.

My hope was to be respected by the person I respected most, and to learn more about him, to deepen our bond by freezing, starving, struggling, and experiencing great danger. My father's dark shadow loomed over my expedition, though I never told him this, and find it hard to admit even to myself. Our journey was an iteration of the oldest story in the world: the son who wants to know his father and be loved by him.

THIS BOOK IS a personal expedition into an area that has been discovered and rediscovered by egocentric, curious people from all over the world, myself included. I am not a historian, theologian, geologist, astronomer, or cartographer, so there is much I don't know about the Pole. My book is therefore not a comprehensive history, but rather concerns the idiosyncratic dreams, ideas, books, and expeditions about the North Pole that have fascinated me for a lifetime.

We are all born with the same instincts as a polar explorer. As soon as we leave our mother's womb, we want more space, more room in which to move. We stretch our arms and legs out in all directions and scream for air. We have a desire to explore the world. As soon as we learn to walk, we walk through the living room and out of the house, then start to wonder what lies between us and the horizon, and soon enough what lies beyond that. We are on our way to discovering our own North Poles.

Our need for more space and adventure will never disappear entirely, as long as we are alive. But today I fear these needs and desires are threatened by mass tourism, and a quick "been there, done that" attitude to life and living. The inherent urge—the dream—to find out for yourself what lies beyond the horizon is somehow being tamed.

I have walked, skied, climbed, and sailed in many parts of the world. I have been able to compare all the mountains, plateaus, forests, plains, and oceans I have seen with somewhere else. But I have only experienced one place that is unlike anywhere else: the North Pole. Because when I finally got there, I realized there was no *there* there.

1

THE FOUR NORTH POLES

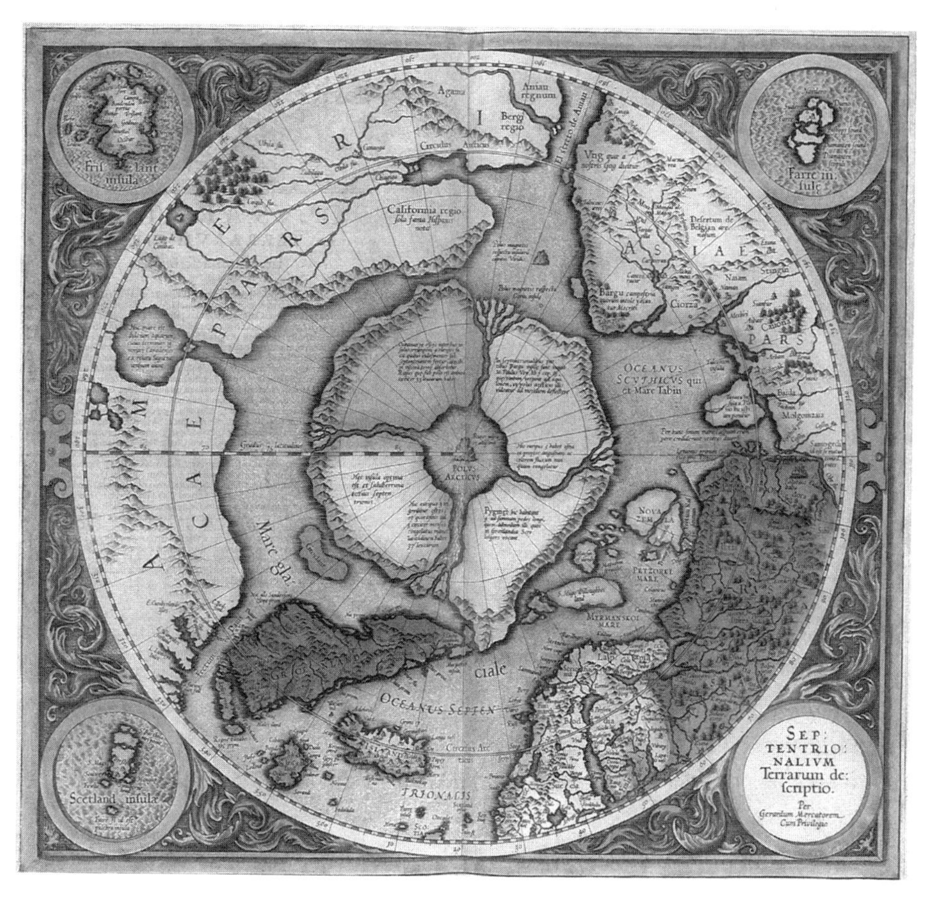

Map of the North Pole from 1606, made by Jodocus Hondius, based on Gerhardus Mercator's groundbreaking 1569 atlas of the world.

A Discovery

On my birthday in 1970 I was given my first globe by my mother and father. It had a metal mount that stretched around the world, from top to bottom. National borders were marked by a thin line, and all other details were shown in different colors that seeped into one another: The oceans, seas, and rivers were blue, glaciers were white, and mountains brown. The bluer the ocean, the deeper it was; the browner the mountains, the higher they were. That little globe helped foster an understanding that nature is connected: water and ice, earth and stone—all are one.

The globe was widest in the middle, where there was a "0" on the mount. From here, the numbers rose and fell to 90. Having studied a large part of the world, my eyes fixed on the uppermost point. What was the large, bluish-white patch on the top? Which country did it belong to? How could you get there?

I was seven years old and had just discovered the North Pole.

I studied the bluish-white area. The North Pole point was hidden under a small chrome disc that secured the globe to its rotational axis, which was attached to the mount. I wondered what was under the disc, what lay hidden at the North Pole. It was thanks to the globe that I realized that the earth spun on its axis: if I turned the globe to the right, it moved east, with the North Pole in the middle. The polar point seemed to be the center of the earth. When my mother told me that at the Equator the earth rotates at a speed of 1,037 miles per hour, I tried to spin the globe as fast as I could. But it was hard to grasp whether I'd hit that speed. Then I had another idea. I stood in the middle of the room, closed my eyes, and tried to spin at 1,037 miles per hour.

IT WAS THE geographic North Pole I discovered when I was seven. The place where the needle on a compass always points south; where the wind always comes from the south because there is nothing more northern than the North Pole. A point where the earth's centrifugal force stops. A point where there is only one sunset and one sunrise each year. The sun sets at the North Pole on September 22, the autumn equinox, and appears again on March 21, the spring equinox.

The geographic North Pole is the point that nations and explorers have vied to reach for centuries, and the point that, on my skis twenty years later, I was aiming for.

It was only years after I got my first globe that I understood that there were in fact three[1] North Poles in addition to the one I would eventually reach. The four North Poles are connected, but each is very different.

The celestial North Pole is located directly above the North Pole, in a straight line, like an invisible continuation of the axis that runs from the South Pole, through earth, to the North Pole and can—with good reason—be seen as the center of the sky.[2]

The magnetic North Pole has been crucial to navigation over land and sea since people realized that a magnet points north. It is a point of survival.

The fourth pole is not so well known, and has never been given a good name. I have chosen to call it the imaginary North Pole. It is not a physical place, but rather is a pole that exists in people's minds, inspired by *eventyrlyst*, marvel, the imagination, and on occasion, both oral and written sources.

The Celestial North Pole

Once I had discovered the geographic North Pole as a boy, I began to wonder what might lie above it. At first, I thought there was nothing, that the earth simply "stopped." But if you stand at the North Pole in the dark of winter and look up 90 degrees to the stars, you look directly into the celestial North Pole.

This is the point in the sky where the earth's rotational axis extends up vertically from the geographic North Pole to the North Star and on into the eternity of the universe. The North Star, or Polaris, is comprised of three stars that appear as one, and is part of the constellation Ursa Minor, also known as Little Bear. The seven brightest stars in the constellation Ursa Major, or Big Bear, make up the Plough.

If you locate the two outermost stars in the Plough, Merak and Dubhe, then draw a line straight up, five times the distance between the two, you will find the North Star. This rule of thumb always works as Ursa Major and Ursa Minor both orbit the North Pole. The North Star, the brightest of the seven stars in Ursa Minor, is directly above the North Pole.

The North Star is vital to navigation: It is a fixed point in the sky, and always in the north. All the other stars orbit this polar point. However, it is not so easy to navigate by the North Star when you are walking to the North Pole as, that far north, it appears to be directly above you the entire time. And in spring, when the polar night is replaced by the midnight sun, it is no longer visible. But farther south, when I sailed across the Atlantic and the Pacific Oceans, the North Star was a constant companion. On my voyage south across the South Pacific Ocean to Antarctica, I felt almost bereft when we crossed the Equator by the Galapagos, and the North Star disappeared for good.

But Polaris has not always been true north. Its position changes slowly over time. Two thousand years ago, the North Star was some 12 degrees away from the celestial North Pole. Kochab, another star in Ursa Minor, was far closer. Astronomers have calculated that over the course of the next 14,000 years Polaris will no longer be the star closest to the celestial pole, and that Vega, a star that is moving toward the celestial North Pole, may become the new North Star.

In Homer, Ursa Major was the constellation that never sank below the horizon and was used as a celestial reference point. When Odysseus is finally about to sail home, the nymph Calypso tells him to keep the Bear, which turns round and round where it is, on his left. If Calypso had tricked him and advised him to keep the sun or the moon on his left instead, he would never have reached home.

Ursa Major is important for those of us who live in the northern hemisphere, as the seven stars are so easy to spot. In the Book of Job, God created the Bear, and in the preamble to the Book of Revelation, Jesus appeared before John in a cave on Patmos in the Aegean Sea, holding seven stars in his right hand.[3]

In Roman mythology, Big Bear and Little Bear were created by Jupiter, king of the gods. Jupiter was married to Juno, but had an affair with a woman called Callisto. When Juno later discovers that Callisto has a son, she assumes her philandering husband is the father. Her revenge is to turn Callisto into a bear so that Jupiter will no longer desire her. Callisto, now living as a bear, is then almost killed when she meets her son, who is on a bear hunt. To avoid catastrophe, Jupiter turns her son into a bear as well, and places them both in the heavens, thus creating Ursa Major (Callisto) and Ursa Minor (her son). This area of sky was called Helike, which means "to turn," as both constellations always closely circle the North Pole.

It was the Greeks who gave the northernmost area of the world its name: Arctic (*Arktos* is Greek for "bear"). They called the opposite end of the earth Antarctic, or "Antibear," an area without bears.

The bear motif is also central to Native American traditions. The Lakota tribe call Ursa Major *Wičhákhiyuhapi*, or "Great Bear." The Mohawk, Seneca Oneida, Onondaga, and Cayuga tribes see the three stars that form a line—Alioth, Mizar, and Alkaid—as three hunters in pursuit of the great bear. Alioth has a bow and arrow to kill the bear, Mizar carries a big pot on his shoulder to cook the meat, and Alkaid is pulling a large bundle of wood to build a fire under the pot.[4]

Why the constellation is called the Bear in several places and on different continents is still a mystery to me. I have never accepted the most common explanation, which is that the Great Bear and the Arctic were given their names because the northerly constellation *looked* like a bear. Or that the area in the south was called the land without bears, because there was no constellation over the South Pole that resembled a bear. Even with the best will in the world, when I look up at the sky, I still don't see that Ursa Major looks like a bear.

I think we are underestimating the intellectual capacity of the people who named the two constellations after bears. People lived close to nature, and I find it hard to believe that Europeans and Native Americans did not know about the brown bear in the northern hemisphere, even if they lived in areas with no bears. It is possible they may also have heard stories about polar bears drifting on ice floes from Greenland to Iceland. And merchants and explorers may have told them about wildlife in the far south and that there were no bears there. If that is the case, the names Ursa Major and Arctic could stem from a knowledge of where bears live, not from imagining a bear in the constellation. (Conversely, Antarctica was given its name because there were no bears there.)

THERE IS NO polar star over the South Pole by which to navigate, or around which the southern stars can orbit. I was the first person to walk alone and unsupported to the South Pole, between November 18, 1992, and January 7, 1993. I saw no sign of life along the way, and used neither radio nor telephone; nor did I have Børge as my companion, so I barely uttered a word. I often looked up at the sky and the celestial South Pole as I walked. But the midnight sun was shining, so I could not see the stars. In March 1990, it was different, as Børge and I were in the darkness for the first weeks as we headed toward the North Pole. We felt we were getting closer to the origins of an ever-expanding universe.

When you raise your eyes and look up at the starry sky you will always be looking back in time. The entire universe is our history, and the speed of light is 186,000 miles per second. Light takes eight minutes to travel from the sun to the earth, a minute from the moon, and a second to orbit the globe eight times at the poles. The light from the North Star is so faint that it is not always possible to see it. When looking for it, I have often thought that it is not as bright as the moon or the sun, but the fact is, the North Star is 323 light-years away, so what we see of it has taken 323 years to reach earth, which takes us back to around the start of the Enlightenment. Back then, the light from the North Star was 4,000 times stronger than the sun.[5]

It's a long way from knowing something to really understanding it. I cannot claim to understand this any better now than I did when I was seven.

The Magnetic North Pole

When we skied to the North Pole in 1990, the needle on my compass did not point true north. The needle is magnetic and pointed to the magnetic North Pole, which at the time I went there was close to Ellef Ringnes Island, one of the largest uninhabited islands in the world, to the west of Greenland. The magnetic North Pole is the point in the northern hemisphere where the magnetic axis pierces the surface of the earth, as opposed to the North Pole, which is where the earth's rotational axis does the same.

Why are the magnetic poles so important? Planet Earth is an enormous magnet, and like all magnets, it has two poles, a plus and a minus. These magnetic fields are essential to all life on earth. Without two magnetic poles, one in the north and one in the south, and the magnetic fields in between, life on earth would cease. When the magnetic fields disappeared on Mars around 3.9 billion years ago, so did the atmosphere and the water.

Magnetism is caused by molten iron in the earth's outer core, roughly 1,800 miles below the surface. Unlike the magnets that you perhaps use to attach photographs to the fridge, the two magnetic poles often shift, due to the movement of the iron in relation to the earth's rotation and its varying temperature. The motion of the iron creates electric currents that produce, or induce, magnetic fields. The ice is therefore not the only thing that is constantly moving in the Arctic—the magnetic North Pole is too. In the thirty years or so since I reached the North Pole, the magnetic North Pole has moved 249 miles north from Ellef Ringnes Island out into the Arctic Ocean.[6] It takes around 300,000 years for the earth's magnetic north and south poles to gradually flip locations.[7]

Because I am more interested than most people in the poles and magnetic fields, I have started to notice how these magnetic fields affect daily life around us. Whenever I go for a walk with a dog that is not on a leash, it often turns around itself before doing its business, as if it is looking for the right direction and checking its inbuilt

GPS. I have found myself wondering why a dog, when left to its own devices, often ends up parallel to the earth's north–south magnetic field. Why is the direction important for the dog?

In 2014, a two-year research project found that dogs choose to "excrete with the body being aligned along the north–south axis under calm magnetic field conditions." This was based on the observation of 1,893 defecations and 5,582 urinations, involving seventy dogs of thirty-two different pedigrees.[8] What researchers have to observe in the name of science!

There has also been extensive research on birds and magnetism. Migratory birds can read magnetic fields, using their eyes, ears, and beak. This is why they can differentiate north from south and, twice a year, fly a quarter of the way round the world and land in a country where summer has just begun. It is quite possible that they do not even know that winter exists. All birds have a protein in their eyes, Cry4, that makes it possible for them to "see" or measure the magnetic fields, day and night. They also have a tiny piece of iron in their inner ear that allows them to feel the magnetic fields, and nerves in their beaks that can estimate the angle to the magnetic fields and thus determine their own position.[9]

Sea turtles and whales use the earth's magnetic fields to navigate with astonishing precision below the surface of the ocean. Bees, whose brains are the size of a sesame seed, find their way back to places where they have previously found food, and home again, thanks to magnetic fields. Livestock, deer, and reindeer like to align with the north–south axis while they graze. Natural fluctuations in the earth's magnetic fields upset all living things, but the sun, stars, and landmarks such as islands and mountains help animals, birds, and humans to correct this, enabling us to find our way.

When I look at a compass needle, which always points north, I sometimes wonder if our relationship with the north is also affected by the pull of the magnetic fields to the North Pole. We have cells that contain iron-based elements, which may influence our blood pressure, pulse, and mood. Scientists at California Institute of Technology believe that our cells can rotate in much the same way that the needle does in a compass, thus sending signals about the cardinal points to our brain.[10] They have discovered that that is certainly true of fruit fly cells, and they maintain that butterflies, rats, whales, and humans are likely to have cells with similar characteristics.

"We are part of Earth's magnetic biosphere," says Joseph Kirschvink, the professor whose research team at California Institute of Technology is working to prove that the "magnetic sense" our ancestors supposedly had several million years ago has not been lost. Kirschvink is not sure what the conclusions will be, but he hopes to prove that the earth's magnetic fields do affect our consciousness and behavior.[11]

I think Kirschvink, or his successors, will succeed. It is hard to imagine that we humans are basically the only creatures that are not influenced by the energy from the north.

The Geographic North Pole

On our way to the North Pole, I never once asked myself why I wanted to walk into a frozen, white-gray-blue nothingness for weeks on end. I was too cold, hungry, and exhausted to ask myself anything. Daily life at home involves endless choices, a constant series of crossroads, whereas life on the way to the North Pole involved only one thing: putting one foot in front of the other.

In purely mechanical terms, it is easy to get to the North Pole:

You simply have to walk far enough in one direction. It lies more or less in the middle of the Arctic Ocean, which spans the northern part of the Arctic region and is surrounded by three continents—Europe, Asia, and North America. It is the smallest of the world's five oceans.

The ice of the North Pole is not fixed to landmass. When you lie down to sleep on the ice at the North Pole, you drift off and wake up in a different place. Once we finally reached the North Pole, we camped there for five days and nights, and in the course of that time, drifted roughly twenty-five miles south toward Ellesmere Island. The opposite is true in Antarctica, a continent surrounded by three oceans—the Pacific, Atlantic, and Indian Oceans. The ice that covers that continent rests on solid ground and is more or less static. When I lay down to sleep at the South Pole, I had moved a mere .13 millimeters north when I woke up eight hours later.

Depending on whether it is spring, summer, autumn, or winter, with significant differences from year to year, the Arctic Ocean is covered by 1.3 million to 6.2 million square miles of ice. The pack ice is made up of ice floes that drift across the Arctic Ocean. Water has a rare property: Its weight in relation to volume diminishes in solid form, so when water freezes, it floats, rather than sinking.

The ice floes move at varying speeds on local currents and the two main ocean currents in the Arctic: the Beaufort Gyre and Transpolar Drift. The pack ice can be anywhere from a couple of centimeters to several feet thick and can drift as much as six to twelve miles within a twenty-four-hour period, depending on the wind and weather. The current between the North Pole and Canada is generally south-flowing; I still remember the feeling of desperation on a number of days when we drifted south at almost the same speed that we progressed north. As we got into our sleeping bags at night,

Børge and I commented that it was a bit like going the wrong way up an enormous escalator, made from ice.

Because the ice is constantly moving, the ice floes press against each other and break up. They are forced up into the air and down into the ocean, creating enormous blocks of ice that fuse together, making them difficult to pass on foot or on skis. These ice ridges are called hummocks.

The American polar explorer Robert Peary (1856–1920) claimed that he had come across hummocks so tall that they reminded him of Capitol Hill in his hometown, Washington. That seems to be an exaggeration; the tallest hummocks that we experienced were around thirty-three feet high, but their height was often not the greatest problem. The ridges sometimes stretched a long way in an east–west direction, and if we were unable to climb over them with the pulks, we were forced to go west or east to where the hummock was passable, which might mean going as far as .6 miles in the wrong direction.

WHEN YOU USE your eyes and a compass to navigate, you have to take the magnetic fields into account, and at the same time ignore them. I knew all this, but still had to take a deep breath the first time I saw the needle pointing west to the magnetic North Pole, on March 8, 1990, 90 degrees off the direction we had to go.

To know what is true north, the direction to the geographic North Pole, and also true south, east, and west, is easy if you understand the compass's deviation. That is to say, the extent to which the compass needle, which points to the magnetic North Pole, deviates from true north. When we skied north, this deviation varied from day to day; the North Pole felt like a moving target.

In the northern hemisphere, the sun is always at its highest above the horizon at midday. And when the sun is at its zenith it is due south to where you are.[12] When the sun is south, the longitude will be 180°. I adjusted the compass every day when we walked toward the North Pole. At six in the morning, the sun is always 90° E, and at six in the evening, it is always 270° W. With every hour that passes, the sun appears to move 15 degrees of longitude west. So every day at 9 a.m. local time, the sun is in the southeast and at 3 p.m. it is in the southwest.

The same principle about longitude and time works in the rest of the world. Sweden is about 15° E, and the time is one hour ahead of Greenwich Mean Time, and New York is almost 75° W and five hours behind GMT.

On the way to the North Pole, if there is no rush, you can find the cardinal points and work out the time without looking at a clock. All you need is to make a primitive sundial: stand your ski pole straight up on the ice or the ground—the shadow from the ski pole will always be shortest when the sun is in the south, because the sun then is at its highest. Before and after midday, local time, when the sun is lower on the horizon, the shadow will be longer, and at its longest around sunrise in the east and sunset in the west.

THE NORTH POLE is both the point from which time all over the world is set, and the point where the twenty-four hours of the day, in relation to the orbits of the earth and sun, are pointless. At both poles, you can decide yourself what time it is. Ever since clocks were synchronized toward the end of the nineteenth century, the world has recognized the significance of the poles as the points that define local time around the world, and yet they are the only places in the world that are not ruled by the clock. The Pole is, in effect, timeless.

When I was twelve, I read the book *Papillon* by Henri Charrière and learned about the arcing path of the sun. The book tells the story of Charrière, a.k.a. Papillon, who was sentenced to life on Devil's Island off French Guiana for a murder he did not commit. Papillon did not have a watch, so during one of his attempts to escape he used a stick to work out the time. French Guiana lies 4° N, and on the Equator the sun is directly overhead at midday. Papillon stood his stick on the ground and waited. When, finally, it cast no shadow, he knew what time it was. I followed him night and day, through all of his eight attempts at freedom, until he eventually managed to flee Devil's Island fourteen years later, on a small raft made from coconuts. I was so fascinated by his tenacity that when I finished the book, I turned to the front so I could begin again.[13]

But it is different at the North Pole. You are no longer in a time zone, your position does not have a longitude and the sun stays at the same height over the horizon for twenty-four hours a day. It moves neither up nor down, but continues west and then east, at 15 degrees per hour, a full circle of 360 degrees. If you stick a ski pole in the snow, the shadow is the same length all day long.

Every time Børge and I left our tent, we followed the path of the sun. The GPS showed that we had reached the Pole, but in the winter of 1990 GPS was still relatively new and presumably no other expeditions had used the technology to reach the North Pole. So, to be completely certain that we were at the polar point, I got out my sextant every few hours to measure the height of the sun above the horizon. I lay on my stomach on the ice and propped myself up on my elbows. As the pack ice is so uneven, it is difficult to see the horizon, but we had an artificial horizon with us: a 4 × 4 inch white container made of hardened plastic. Robert Peary writes that he filled a container with mercury, then lay on his stomach on a fur to measure the height of the sun in relation to the flat surface in the container. Mercury has no other use, so I used the purified petrol we were already carrying to minimize the weight.[14] We could pour it back into the bottle once we had used it. And if we ran short of fuel, we could use it on the Primus stove.

I asked a friend who went to the North Pole two years later, in 1992, how he experienced being able to finally follow the sun's straight path along the horizon. He looked at me, baffled. He had been so obsessed with his GPS—which he trusted 100 percent—that he had not looked up at the sun.

I FIRST BECAME aware of the expression "polar explorer" a year or two after I had been given that globe by my parents. It was summer, and I was in a cemetery where even the old graves were well tended. The grass was full and green, and there were flowers blooming in front of the headstones. I was with my great-aunt Tove, who had the inspired idea to read the headstone inscriptions out loud to me. "SAILOR" . . . "CAPTAIN" . . . "JUDGE" . . . and then black letters on a modest gray rock: "POLAR EXPLORER." I didn't know what it meant. Tove explained that it was someone who went to the North or South Pole—and from that headstone the distant dream was sown of becoming a polar explorer. There was something about the respect that the two words engendered. Polar explorer. The thought that the man who lay buried there had traveled so far that he had earned this respect, and his feat—his fate—could be summed up and eternalized in two words.[15]

I HAVE ALWAYS been something of a dreamer. In my imagination, I sailed throughout my childhood with the Norwegian explorer Thor Heyerdahl (1914–2002), on his voyages on *Kon-Tiki* over the Pacific Ocean, on *Tigris* over the Indian Ocean, and *Ra* over the Atlantic. I skied with Fridtjof Nansen (1861–1930) and swung on the vine-like lianas with Tarzan, was marooned with Robinson Crusoe and escaped Devil's Island with Papillon. When I was due to start school, my parents were advised to keep me back a year, so I started primary school when I was seven and a half. I think they thought I was immature, and they were probably right. I was also dyslexic, which meant that I did not learn to read until I was ten. Fortunately, my parents read to me. My father read Tarzan, and my mother fed me literature, particularly Homer. Odysseus became my hero when he overcame the urge to abandon himself to the sirens' bewitching song and sailed past their island of Anthemoessa. Nothing seemed impossible. Dreams were more exciting than daily life and I found it hard to separate reality from imagination. Admittedly, this had both advantages and disadvantages. I thought I was different but, as I got older, I realized that most people are the heroes of their own dream life.[16]

Tove, the great-aunt who introduced me to the notion of polar explorers, had known the famous Thor Heyerdahl. She told me that he had nearly drowned as a boy and was thus afraid of water. This nugget of information has given me much pleasure ever since. I slowly came to understand that explorers travel despite the fear, not because they feel no fear. Soon after I had learned to read, Tove lent me a book about the Portuguese explorer Ferdinand Magellan (1480–1521) and the great turning points in history—those fateful decisions that have far-reaching consequences beyond their time. It was called *Stillehavets beseirer* and has never been translated into English. But had it been, the title would have been *Conquerors of the Pacific*.

I still remember the story of Magellan's discovery of the sea route from the Atlantic Ocean to the ocean he named the Pacific.[17] Magellan's chronicler, Antonio Pigafetta (*c.* 1491–1531), explained in his diary that it was given this name "in the hope that it would remain peaceful." The strait that Magellan sailed through was named after him and lies between mainland South America and Tierra del Fuego. I have sailed it myself, and it is narrow in places, with difficult shallows, tides, islands, and unpredictable winds. The key difference between my voyage and Magellan's is that we had a chart that allowed us to navigate safely between the islands and shallows, and we knew that there was open ocean at the other end. In Magellan's day, there were people living on the east side of the strait, who presumably gave the crew valuable information about what to expect. It still took Magellan a month to navigate the strait, and us a mere forty-eight hours.

Magellan believed that the ship would reach land within a matter of weeks when they set out across the Pacific Ocean, but they sailed west for three months and twenty days. Like Christopher Columbus (1451–1506), Magellan thought the world was smaller than it is. They were eventually forced to eat the rats on board the ship: "Rats were sold for one and a half ducats apiece, if anyone managed to catch one." I imagined the men in rags, starving, running and crawling on their bellies in pursuit of the rats, skinning them, frying them, and eating the meat, which I could only assume was stringy. A minor mystery that science had still not unraveled at the time was why rats did not get scurvy. We now know that rats, unlike humans, produce vitamin C themselves. Without knowing it, eating the critters provided the crew with the necessary vitamins and thus reduced the dangers of the deadly disease.[18] Many years after I had sailed through the strait, I was offered rat meat in a café in Togo. I thought about Magellan, and accepted.

Magellan was beaten to death on the island of Mactan, in what is now called the Philippines. The natives, who the Europeans claimed to have discovered, interpreted Magellan's arrival as an act of war. However, one of his five ships, *Victoria*, made it all the way across the Indian Ocean, rounded the Cape of Good Hope, and sailed back to Spain, thus completing the first circumnavigation of the world. The ships that originally set off had about 270 men aboard; the one that returned had eighteen. Pigafetta was one of them. If he had died en route, we would never have had his account of the journey.

A surprise awaited the crew on their return. According to the logbook, they had been away a day longer than the number of days since they'd recorded that they left Spain. No one on board knew that a natural consequence of sailing west around the world, following the sun, was that it would take them a day further into the future.

When I was reading this book, it was of no importance to me who had undertaken the voyage and from where—what mattered was the incredible drama of the adventure. The story itself inspired me. I wanted to sail across oceans as well, a dream that was fulfilled nine years later, in the autumn of 1983, when I crossed the Atlantic Ocean to the Caribbean with three friends.[19] It was no straightforward trip: On our way back to Norway, in the winter of 1984, our thirty-five-foot sailing boat almost sank during a storm in the North Atlantic. We had four "knockdowns"—the top of the mast hitting the ocean—in a single day; the boat was filled with salt water; the engine didn't work; nor did the toilet. The entire trip was one long, brutal lesson in the importance of being better prepared. I was soaking wet for the twelve days following the knockdowns, got frostbite, and before we reached Norway, decided never again to expose myself to so much cold weather.

THE OLDEST ACCOUNTS of people traveling out into the world, and many of the books I have read about polar expeditions, are often similar to Magellan's story. Heroes—or heroes in waiting—travel to mysterious places and either die there or come home to tell the tale. There is every reason to claim, as the American author Paul Zweig does in his book *The Adventurer: The Fate of Adventure in the Western World* (1974), "that the narrative art itself arose from the need to tell an adventure; that man risking his life in perilous encounters constitutes the original definition of what is worth talking about."[20]

Zweig refers to the *Epic of Gilgamesh*, perhaps the oldest written story in the world. Gilgamesh lived in Babylon and took the initiative to build a city wall that once was considered one of the world's seven wonders, but then felt hemmed in and sick.[21] He had a restless heart and chose to travel the world, to risk his life in battle with monsters and dragons. He wanted to be admired and famous, and to discover the secret to immortality. Gilgamesh finds the plant that guarantees him eternal life, but a snake eats it while he is asleep, which turns out to be a blessing. One fundamental disadvantage of immortality would surely be that daily life becomes meaningless when death is removed from the equation. Equally, it would be meaningless to try to reach the North Pole if there were no danger to one's life involved. This does not mean that explorers have a death wish, but rather that when there is real danger one feels truly and completely alive.

TOWARD THE END of the 1980s, there were three Norwegian lads who dreamed of reaching the top of the world: Geir Randby, Børge Ousland, and myself. We all had dreams of skiing to the North Pole, but Geir decided on 1990 as the year in which our expedition would start. There is something magical about numbers and dates and expeditions, as it is only when an actual date is set that the plan becomes not just a pipe dream, but a dream with the hope of becoming reality. Once Geir had decided on March 1990 and our preparations had begun, I realized my dreams were flawed: They did not include even the remotest possibility of failure. A challenge is only meaningful when there are dangers—and the chance of failure. If it were not dangerous or difficult, many more people would go to the North Pole.

Geir, unfortunately, slipped a disc in his back ten days into the expedition, when the pulk he was pulling fell down a hummock. He had to give up and was flown out after waiting five days on the ice. Geir was integral to our preparations and was there for the coldest part of the expedition. I felt sorry for him: He didn't have the dose of luck we all need to succeed. Once he was back in Norway, Geir felt he had nothing to live for. A year later, he felt such pain whenever he thought of the expedition that he could not move at all. It seems that it was far harder to give up the dream of reaching the North Pole than it was for us to carry on. But perhaps Geir's is the greater feat, as he overcame the disappointment and carried on exploring the Arctic.

The Imaginary North Pole

Only after we reached the North Pole did I realize that humans in Europe, Asia, and North America had been visiting the imaginary North Pole for thousands of years. I had been obsessed with getting there. For more than two years every book I read, and almost every conversation I had, was about preparing for our expedition. Once I was back in Norway and my obsession had started to subside, I understood that there was so much more to the North Pole than just getting there.

Its history can, broadly, be divided into three parts. The first is the longest: It stretches from the dawn of time until the sixteenth century. During this period, people in the northern hemisphere speculated about what was to be discovered north of the furthest horizon, but were unable to travel there, so it existed purely in our imaginations. In the second part, the 400 years or so from 1500, the ice and cold killed many of those who tried to travel to the Pole in person. And in the third part, through the twentieth century until the present day, the North Pole was finally reached. But according to leading scientists, the ice is now in danger of disappearing in the summer, due to higher temperatures brought on by our reckless behavior.

THROUGHOUT HUMAN HISTORY, people have wondered how the world fits together. Copernicus did not base his theories on substantially better observations than his predecessors—the telescope had yet to be invented—but he and his contemporaries used these existing observations and a rational approach to develop groundbreaking theories.

The Greek astronomer and cartographer Anaximander (610–546 BC) is credited with having made one of the first maps of the world, although without the North Pole. He managed to convince his contemporaries of two revolutionary ideas: not only that the world floated freely in space and did not rest on something physical, but also that the surface of the earth curved, and thus the sky continues under our feet.[22] Anaximander reached both these conclusions more than 2,000 years before Ferdinand Magellan's ship *Victoria* had sailed around the world, and 2,500 years before astronauts had photographed the earth floating in space.

Of course, the results of human speculation are not necessarily accurate, and the earliest explorations of the North Pole are a good example of this. From the 1500s to the 1800s, men tried to picture the North Pole, but never fully imagined the ocean currents and cold, the darkness and the ice, that would greet them. Perhaps they were too confident. Perhaps they lacked Anaximander's curiosity.

Rigveda: The Earliest Source

The earliest mention of the North Pole is in the Vedas, the oldest Hindu scripture. *Veda* means "knowledge" and "insight" in Sanskrit, and the work is comprised of four collections: the Rigveda, the Yajurveda, the Samaveda, and the Atharvaveda. The content evolved over several thousand years, and encompasses myths, teachings, and stories. The Vedas were passed down orally through the generations, until they were written down in the north of India. No one knows the origins of the Vedas, but the work to write them down started around 1500 BC.

The first leader of the Indian independence movement, Lokmanya Bal Gangadhar Tilak (1856–1920), was both a teacher and scholar of the Vedas. *Lokmanya* is an honorary title that means "accepted by the people as their leader." The British colonial powers saw him in a different light, and called him the father of Indian unrest. They imprisoned him in the 1890s, and it was in prison that he finally had the time to study the significance of the North Pole in the Vedas.

In the oldest Veda, Rigveda, Tilak found evidence to argue that the North Pole was the home of the Vedas, as well as of humanity.[23]

The texts he refers to were written by people who studied the night sky and speculated about the celestial North Pole. Sources may also have included people that had been to the Arctic, but it is unlikely they ever reached the North Pole, given it was in the middle of an ocean covered by ice.

In his book *The Arctic Home in the Vedas*, which was published after his time in prison, Tilak writes that the climate at the North Pole was warmer prior to the last ice age, and he is right. When a new ice age took hold, the inhabitants migrated south and populated Europe and parts of Asia. On the basis that islands and continents drift, he surmised that between the last two ice ages, there was mainland at the North Pole, or slightly farther south, but still north of Europe and Asia.

Tilak was not alone in this. Some years later, in 1906, Robert Peary claimed to have seen land in the Arctic Ocean, north of Ellesmere Island, and named it Crocker Land. For some years, the area was called "The Lost Atlantis of the North" and was drawn on new maps.

As late as 1926, the Norwegian polar explorer Roald Amundsen (1872–1928) wrote in the Norwegian newspaper *Arbeiderbladet* that he fantasized about discovering a civilization far up in the Arctic Ocean. "I have often wondered what might have happened if one of the hundreds of whaling ships that get stuck in the ice in the Bering Straits . . . was pulled along by the drifting ice until it hit a small island." There could be as many as fifty men and women on board, a mix of Inuits and Western people. They might have used the ship timbers to build houses and enjoyed the milder summer months. "They could have continued to live there, procreated and been happy," forever cut off from the rest of the world by the drifting ice. Amundsen realized that the idea might seem ridiculous "but it is something I have often wondered about."[24] Amundsen still nursed the hope of finding an island or new country when he flew over the North Pole from Svalbard to Alaska in 1926.

In the Rigveda, the firmament is described as being held up from the earth by a pole, not suspended in space, and the stars move in a circle like a great wheel around the Pole. The stars neither set nor rise, but lower toward the horizon only to ascend again. And the creator of the universe lives beyond the seven stars of Ursa Major.[25]

Mount Meru is the center of all universes in Hindu and Buddhist cosmology. Not only the physical universes, but also the metaphysical and spiritual. No one knows for certain where Meru is located, but Tilak studied the Rigveda and found details that he believed proved that Meru was to be found at the North Pole.[26]

In Hindu mythology, Mount Meru is the center of everything; it is located at the top of the world's axis and is home to the gods. The sun, stars, and moon move around Meru, from west to east, and "the day and the night are together equal to a year to the residents of the place," like it is around the North Pole.[27] There is a description of what could be the Northern Lights: "The mountain, by its lustre, so overcomes the darkness of night."

The Iranian Vendidad, like the Rigveda, talks of a region where a single day and night last an entire year. "There the stars, the moon and the sun are only once a year seen to rise and set, and that year seems only as a day."[28] Tilak argues that as the same description is to be found in both Iranian and Indian teachings, the original source must have existed at a time when the North Pole was inhabited.

In the Arctic, there has also been a long tradition of stories that have survived intact over vast periods of time. The Dano-Greenlandic explorer Knud Rasmussen (1879–1933), who crisscrossed 18,000 miles of Greenland, Arctic Canada, and Alaska by dogsled, wrote about his encounter with a girl on Greenland who recognized a story told by a grandfather in north Alaska, just as the man knew about the traditions of the Inuits in northwest Greenland.

When we were preparing for our expedition to the North Pole in 1990, we spent a month in Iqaluit, the largest town on Baffin Island, in the northern territory of Canada. There I heard a story that confirmed everything I had learned about the importance of writing something down so it can be remembered. The American polar explorer Charles Francis Hall (*c.* 1821–71), visited Baffin in the 1860s. He was one of the first who wanted to learn from the Inuits. To Hall's astonishment, the local Inuits gave him a detailed description of the arrival of the British polar explorer Martin Frobisher with his ships and crew some 300 years earlier.

It was around this time that the Canadian authorities started to "civilize" the Inuits, a campaign that peaked in the mid-twentieth century. Young Inuits were sent to boarding school far from home, where they were taught a curriculum chosen by the Canadian government, and their oral history and culture were almost completely wiped out and replaced by Canadian history and culture.[29] It was a slow but focused form of erasure.

TILAK DISCUSSES USHA, the goddess of dawn, who is mentioned several times in the Rigveda. Sometimes she is talked about in the singular, and at other times in the plural, which Tilak finds strange.[30] Having experienced sunrise over the Arctic Ocean every morning, until there is a midnight sun, it seems perfectly natural to talk about sunrise in these fluid terms. The sunrise at the North Pole is unlike any other. The sun only rises once a year, but further south in the Arctic Ocean, it seems to happen incrementally; the stages are longer each day until there is midnight sun. Dawn goes on and on until the red sun finally appears and creeps along the horizon, then disappears again, and another long twilight begins. Morning after morning the sunrise felt never-ending. Had we been at the North Pole, we would have experienced a continuous sunrise lasting several weeks, slow and languid, until the first sunbeams heralded the sun's arrival over the horizon.[31]

Herodotus and a Civilization at the North Pole

The historian Herodotus's (484–425 BC) principal work, *Histories*, is the oldest existing attempt to write the history of the world. Herodotus's world consisted primarily of the countries around the Mediterranean and North Africa. He also writes about a group of people outside the known world, whom he believed lived in the vicinity of the North Pole. These people were Hyperboreans, a people who, according to the historian, lived north of the north winds, as far from the Greek world as one could possibly go, where there was twenty-four-hour daylight (*hyper* means "beyond" and *boreas* means "the north wind" in Greek). In Greek mythology, Boreas was the god who ruled the cold winds from the north. Like the wind, he was strong and irascible and had wings.

According to Herodotus, no living person had met a Hyperborean, and no one had been to Hyperborea, the land where they lived. All the same, the historian believed there was reason to claim that the people in Hyperborea, due to their position in the world, "possess the things which by us are thought to be the most beautiful and most rare."[32]

Herodotus's Hyperborea was a heavenly paradise where people lived in peace and harmony. All countries went to war from time to time, but not Hyperborea, where eternal peace reigned. Herodotus writes that the Hyperboreans had been in contact with civilization farther south, but had been disappointed by the people there. Two girls, Hyperoche and Laodice, set off on a journey south to our world but never returned home. If tragedy struck once, as it must have for the two girls, then it could happen again. The Hyperboreans realized that it was better not to associate with people from our world and therefore chose to live in isolation at the North Pole.

It is impossible to know today how the idea of Hyperborea was conceived. Herodotus writes that others before him had described the country in the north, but that the texts had been lost.

For the greater part of antiquity, it was usual to think of Hyperborea as a real country, beyond the mountains, where people were happy and lived for a thousand years, where there was only one sunrise and one sunset each year. There are also other accounts of countries to the north of the most northerly mountains, but the myth of Hyperborea has proved to be most resilient. As late as the Enlightenment, Hyperborea was drawn on maps at the top of the world.[33]

Aristotle, Pytheas, and the Art of Exploration

In the century after Herodotus completed his *Histories*, Aristotle wrote his *Metaphysics*. He argued that the pursuit of knowledge is part of human nature; we should explore the ever-changing world around us ourselves and not leave it up to others. But beyond exploration, he continues, reflection is the seed of all knowledge, not just about you and me, but about the world and the entire universe. We initially wonder about what it is we see, smell, hear, feel, and taste, then about what lies beyond the horizon, and everything our senses cannot fathom. I remember that growing curiosity from my own childhood. To wonder and seek knowledge is not only a good idea; it is our moral responsibility. Aristotle pointed out that we can experience the same thing in different ways, so we should work together and share our experiences.

In the 320s BC, a decade after Aristotle had set out an ideological and moral basis for exploration in his *Metaphysics*, the Greek astronomer Pytheas left his hometown of Massalia (now Marseilles) and traveled north.

Polar explorers, late at night in their tents, sometimes talk about which historical expedition they would most like to have been part of. Those of the Englishman William Parry, the Norwegian Fridtjof Nansen, or the American Robert Peary? Each had his merits, but I would choose Pytheas's lesser-known voyage.

It is no longer known how Pytheas might have imagined the world to the north. Educated people in Massalia, like Pytheas, would have been familiar with Homer and the Roman poet Ovid's descriptions of the stars, countries, and weather in the north, as well as Herodotus's description of the Hyperboreans in his *Histories*. In *The Odyssey*, Homer wrote about a mysterious people who lived in a country far away that was shrouded in mist and clouds, a place where the sun never shines. And in *Metamorphosis*, Ovid writes about a dark and silent country that lies to the north of the north wind. According to Ovid, it was the home of Somnus, the god of sleep.[34]

Like other Greek towns on the Mediterranean, Massalia looked out on the water. We cannot be certain what Pytheas knew about what awaited him in the north, but I believe he must have learned something from the experience of others. Most travelers, no matter where they go, like to talk about their travels. Given there were merchants who imported tin from England and amber from the Bay of Bothnia and Denmark to Massalia, there were probably plenty of accounts of the landscapes, people, and cultures in the north.

And he no doubt knew about some of the great voyages of discovery before his time. The explorer Himilco, from Carthage (in what is now Tunisia), was famous in Pytheas's day. Around 500 BC, he sailed around Gibraltar and up the west coast of Europe to the British Isles. On his return, he told tales of terrible sea monsters and seagrass so thick that the boat became entangled in its grip. (Supposedly this was to frighten off other explorers and merchants from exploring these areas.) But his travel account may also have been true. If you were to see a whale for the first time, it would easily match your mental image of a sea monster. And I have seen vast patches of orange sargassum seaweed farther west in the Atlantic. The seaweed was so dense that it was hard to navigate our sailing boat.

No one knows why Pytheas traveled north, other than for adventure. There were probably several reasons: He was perhaps sent north on behalf of his hometown. In which case, the driving force may have been the same as for many later polar explorers, the first landing on the moon, and the race to conquer Mars: prestige, both personal and national. He may also have been looking for new trading opportunities, and, as he was a discerning astronomer, the opportunity to observe planets and stars from a new perspective may also have been an incentive. The need to chart the world was pressing. Pytheas lived at a time when astronomers could measure latitudes (how far north or south you are), but not longitudes (how far east or west you are).[35] If you know the date and look at the sun, it is relatively simple to find the latitude by the length of the shadow cast by the sun.

When Pytheas returned, he wrote a book about the journey, *On the Ocean*, and many were shocked by what he claimed to have seen in the north, and accused him of lying. The book was lost around 2,000 years ago, so what is known of the content is pieced together from those who wrote about it. Nevertheless, the voyage was of great importance to astronomers, cartographers, and geographers in his day and the centuries that followed. It fired an interest in the north at a time when the geographic North Pole was yet to be positioned on a globe or map. In his time, he was the only source in the Mediterranean with firsthand experience from the far north.[36]

I have sailed much the same route from Marseilles to Norway. The Mediterranean, Bay of Biscay, and North Sea are now well known for their strong winds and rough seas, but at the time Pytheas was sailing more or less into the unknown.

After he had sailed past the British Isles, Pytheas continued north for six days and six nights before he reached land again.[37] No one knows for certain where exactly this was. It could not have been the Orkneys or Shetland, as they lie only a day or two north of the mainland. It was most probably Norway or Iceland, but it was unlikely that Iceland was populated at the time. Pytheas described a land where the sun only dipped below the horizon for a few hours a night and it never got properly dark in the summer. The local inhabitants advised Pytheas on where he should sail—to the place where the sun goes to rest.[38]

Even farther north, he reported that the sun did not go down at all in summer. He saw the Northern Lights and exchanged stories with the local people he met along the way. Aristotle had said that Athens lay at the center of the civilized world, and that those who lived in the center lived the good life. The further away you lived, the less civilized you were. And it was too cold in the far north for anyone to live the good life. Pytheas would have been able to verify whether Aristotle's theory was true or not. Fishing, farming, and trade were well developed along the Norwegian coast at the time, so I like to think that when he finally returned to Massalia, he was able to inform his compatriots that Aristotle had been entirely wrong on this count.

Pytheas tried to continue sailing north, and the ancient Greek geographer and historian Strabo says that he described a reality where land, sea, and air ceased to exist separately, but merged as one entity. The way ahead was neither passable nor navigable. These surroundings could have been ice slush, or an area of thick fog where the water had frozen. The ocean was transformed into congealed sea, an unpassable stiff mass of cold water and air. I am reminded of my own experience when thick fog lay like a cold, damp cloth over the ice. He compares the ice floes he saw with a huge jellyfish and was ridiculed for it. Yet both appear as a whitish-gray mass floating on the surface. And besides, it is not easy to describe what an ice floe looks like, especially to somebody who has never seen one.

To be able to experience such nature before almost anyone else is why I would, of all historical expeditions, have chosen to join Pytheas.

In the centuries that followed, both historians and geologists claimed variously that the ocean in the north does not freeze, that no one lives north of the British Isles, and that Pytheas had made it all up. Strabo called him the greatest liar in history. But there were also scholars who supported him. The pleasing irony is that much of what Strabo repeatedly said was lies has turned out to be true. Pytheas called the most northerly known point on earth Thule. He never went there himself but said that it was a place with perpetual light. The word *Ultima* was added later in antiquity. Ultima Thula was, like Hyperborea, an idea—a terra incognita—of what lay north of the limits of the known world.

WHEN SITTING AT home, it is all too easy to come up with good reasons not to do anything, but when you are out in the wilds, you become more single-minded and think less about sensible reasons for doing what you are doing. You feel that you are part of the ocean, the wind, and the sun. Adventure is all about deliberately making life more difficult than it needs to be, and having far less control over yourself and your surroundings.

I think that, like me, the longing for adventure and the curiosity to know what lay beyond the horizon was so strong that Pytheas tried to get as far north as he could. It was that same longing that drove me to the North Pole more than two thousand years later.

Alexandria: The Gateway to the North Pole

Today it is a given that the North Pole is at the top of the globe, but for a long time this was far from clear. Historically, there have been two centers of learning that decided how the world should be drawn. One was in what is now the Netherlands and Belgium. It was here that modern cartography was established in the sixteenth century. The position of the North Pole has been more or less undisputed since 1569.

But before that, nearly 2,000 years earlier, the Egyptian port of Alexandria, some 4,060 miles from the North Pole, was the leading center of astronomy, cartography, and geography. And it was here that the foundations were laid which the Europeans developed later.

Alexandria was a city like no other. One of the seven wonders of the ancient world, a more than three-hundred-foot-tall stone lighthouse stood by the port. A mirror at the top of the tower reflected the sun out to sea during the day, and at night a fire was lit so travelers could more easily find their way. Countless ships and caravans visited the city, and travelers exchanged stories about people, animals, climate, trade, and navigation. All maps and books that came into the city were copied and the copies were stored in the library. The library is said to have housed some 100,000 manuscripts.

But in order to establish the position of the North Pole, it was first necessary to answer three fundamental questions.

First, it had to be proven that the world was round. Then the circumference had to be calculated. It was not possible to determine the position of the North Pole before the size of the earth was known. Scholars then also had to agree that there were two poles, and that the North Pole would be at the top and the South Pole would be at the bottom.

As early as the fifth century BC, it was common knowledge in Alexandria that the world was round. When ships left the port and sailed out toward the horizon, the inhabitants observed what anyone who has seen a boat leave port has witnessed. First the boat disappears over the horizon and then the mast. And when the boats returned, people saw the masts first and then the boat. There was little doubt that the earth curved. This theory was confirmed around 350 BC when Aristotle (384–322 BC) supposedly observed the circular shadow of the earth on the moon during a solar eclipse.

The astronomer, geographer, and mathematician Eratosthenes (276–194 BC) was the chief librarian in Alexandria and is credited with coining the word *geography* (*geo* means "world" or "earth," and *grafi* "to write" or "to describe"). It was he who decided to measure the circumference of the earth. Eratosthenes was familiar with Pytheas's measurement of the latitudes when the adventurer voyaged north and used these as his foundations. With no knowledge of the North Pole, he drew a map of the world and placed an island—Thule—farthest north.

He calculated the circumference of the earth by comparing the height of the sun above the horizon in two places, Alexandria and what is now Aswan in the far south of Egypt, on the summer solstice. The distance between the two places was then carefully paced and measured. Eratosthenes compared the distance with the sun's position at midday to calculate the circumference of the earth. He concluded that the circumference was somewhere between 25,010 and 28,520 miles. He wasn't far off: The earth's circumference at the Equator is 24,901 miles.

This theory about the size of the earth caused alarm in late antiquity—and on into the Middle Ages. The world was far larger than had previously been assumed, and the size indicated that there might be people living on the other side of the world. For many this was unimaginable, and later, in the Middle Ages, the Catholic Church rejected the idea that God might have created people and countries that were not mentioned in the Bible.

One of the Church fathers, St. Augustine of Hippo (AD 354–430), wrote that there could not possibly be people where the sun shone after it had set in the known world for one obvious reason. They were too far away to have possibly been descended from Adam, and the Bible was not false.[39] Only when America was discovered, with its unknown people and animals, did the Church change its worldview.

OVER THE CENTURIES, it has not only been the servants of God who mistrusted Eratosthenes's teaching. One such doubter, 1,700 years later, was Christopher Columbus. He believed the world was a third smaller, an error that was naturally of great significance to his four voyages to America. Columbus was well versed in geography, astronomy, and history, and everyone in the circles in which he moved knew the world was round.

But unlike many of his contemporaries, Columbus actually believed that the world was more pear-shaped. In a letter to Queen Isabella and King Ferdinand of Spain, he explains that the greater part of the world is round like a ball, but the top is pointed, like a pear's stalk. The upper half of the earth can therefore be compared to a woman's breast and the top, to quote Columbus, is "like a woman's nipple." It is the highest point in the world and closest to heaven, but it lies to the east, not to the north. He calls the top "the end of the East, where end all the land and isles."[40] When Columbus sailed west, he hoped he would get so far west that he would sail around the world, into the eastern part of the world, and reach the top, where the Garden of Eden, the paradise he knew from the Bible, had once been.

THOR HEYERDAHL ONCE humorously said that it was important for explorers to have the right wife at the right time, which Columbus did. His first father-in-law sailed extensively on the eastern side of the Atlantic Ocean, and may well have told his son-in-law about the trade winds. I have sailed roughly the same route as Columbus over the Atlantic: We headed south from the Iberian Peninsula to catch the trade wind from the Canaries and Cape Verde across to the Caribbean. It is a northeasterly and, if you catch it, you will have a comfortable run nearly all the way. On his first voyage, Columbus headed west too early, but when I look at the map, it seems that he sailed farther south on his third crossing and managed to catch the same trade wind as we had. It is the perfect route for getting to the Caribbean—but not Asia. If Columbus had navigated according to Eratosthenes's calculations, rather than his own belief in a smaller world, he would have sailed farther south and perhaps never got to America.

THE BABYLONIAN WORLD map is deemed to be the oldest map of the lands, rivers, and oceans known to any cartographer. Yet the North Pole is conspicuously absent. The map is drawn on a clay tablet and dates from between 800 and 600 BC, and the world is surrounded by two circles that contain salt water. The Euphrates River is at the center and runs in a north–south direction, with the city of Babylon on its banks. North is at the top of the map, and south to the bottom. No one knows why, but it is generally thought that observations of the celestial North Pole, a fixed point in the sky, far north of the map's most northerly boundary, have influenced the makers of the map.

The first recorded time that the North Pole was positioned at the top of the world was on a globe that no longer exists. The globe is thought to have been made by the cartographer, librarian, and philosopher Crates of Mallus in 150 BC. Crates was an expert on Homer and used *The Odyssey* and Pytheas's travel accounts as a basis from which to understand astronomy and the geography of the world.

Some 300 years later, according to one theory, around AD 150, the North Pole was placed at the top of a world map. The cartographer, astronomer, and mathematician Ptolemy of Alexandria (AD 100–170) concluded that this was the correct position for the North Pole. It is hard to understand why it should take 300 years for this to happen. Cartographers other than Crates and Ptolemy must surely have come to the same conclusion, only to then be forgotten.

Ptolemy's great innovation was that he used longitudes and latitudes to establish where on the map towns, mountains, oceans, and the North Pole should be. But why the North Pole ended up at the top of the map remains an open question.

Ptolemy barely left Egypt during his lifetime and he never traveled north to find out what was hidden under the celestial North Pole. But he did have a major advantage over earlier scholars and those who lived elsewhere: Much of the Alexandrian library was still intact and the city was still home to leading experts in mathematics and astronomy. He could build on the knowledge of Anaximander, Pytheas, and Eratosthenes.

THE USUAL ANSWER to the question of why Ptolemy positioned the North Pole at the top of the world is that the inhabitants of Alexandria knew more about the north than the south.

Ptolemy and his Alexandrian contemporaries spent much of their time trying to learn from the stars. They stayed up at night and studied their movement around the North Pole, from east to west, in twenty-four-hour cycles. In Egypt, an individual might be a specialist in many fields of knowledge: priest, cartographer, astrologer, and astronomer all at once. It was commonplace to think there was wisdom hidden in the stars.

It is entirely healthy to let your eyes rest on the stars to gain a deeper understanding of oneself, to recognize that feeling of humility, a reminder that you are but a tiny part of the endless universe. A speck of greatness because you are, after all, a living part of the universe. Polar explorer Fridtjof Nansen wrote about the sky at night: "It is ever there, it gives ever peace, and reminds you that your restlessness, your doubt, your pains are passing trivialities.

The universe is and will remain steadfast. Our opinions, our struggles, or sufferings are not so important and unique, when all is said and done."[41]

THE GREAT PYRAMID of Giza outside Cairo was built as a burial chamber from 2580 to 2540 BC. Its shape and location were chosen to reflect the star-strewn sky and give the cosmos a form on earth. With its 2.3 million blocks of limestone, it is the heaviest building ever made. The sarcophagus was positioned on the west side of the pyramid. There are two doors marked on the stone wall facing east, one to give a view to the sunrise, the other to give a view to the world. There are also shafts through the pyramid that align with Orion and Sirius. The architectural calculations were based on what was known about the sun, planets, and stars. If you visit and have a compass with you, as I had, you will see that the cardinal points on which the construction of the Great Pyramid is based are more precise than any compass. A compass deviates some 4–5 degrees as a result of the earth's magnetic field, but thanks to the Egyptians' own observations their calculations were considerably more accurate. The entrance to the Great Pyramid faces, effectively, straight north. The enormous building is one twentieth of a degree, or eight inches, off true north.

The north was important, as after his death the king's soul would fly through the pyramid and on to what we now call the celestial North Pole. This journey would start in a straight line through a north-facing shaft of light. The choice of this northerly position in the sky was not arbitrary. The stars that orbit around the polar point and never disappear below the horizon, such as Dubhe in Ursa Major and Kochab in Ursa Minor, symbolized eternal life. There was no text inside the burial chambers in earlier pyramids, but in the later pyramids there are hieroglyphs from floor to ceiling, explaining life after death, and the journey to the celestial North Pole.[42] "I [the king] will cross to that side on which are the Imperishable Stars, that I may be among them."[43]

Nowadays, you could live your whole life in Cairo, only twelve miles from the Great Pyramid, or in any other big city, and never see the stars. There is so much artificial light in the city that the stars are no longer visible in the night sky.[44] When I walked around the streets at night, after my visit to the Great Pyramid, it felt like a small tragedy that a natural source of wisdom was gone, and most of the city's ten million inhabitants live as though the stars no longer exist.

The First People on Earth

The idea that the North Pole is the world's lost paradise, where the first humans lived, is an old one, though its origins are unknown. The theory spread across the greater part of the world and has proved remarkably resilient.

As late as 1885, Professor William F. Warren, the first president of Boston University, claimed in his book *Paradise Found* that Adam and Eve had lived at the North Pole.[45] The top of the world was "the cradle of the human race," before a change in the climate caused global cooling and people had to migrate south.[46]

Warren writes that while many of the world's great mysteries had been solved in previous centuries, the most important question for historians, paleontologists, and archaeologists remained unanswered, namely: Where did the first humans live?

Explorers had, according to Warren, ventured out into the world with this question in mind, but none had come back with the answer.

He writes that it was naive of the missionary and explorer David Livingstone to travel to the source of the Nile in Africa to find the answer to the origins of man. But he is even more scathing about Christopher Columbus: He goes as far as to say that Columbus's account of his third, and penultimate, voyage over the Atlantic for his royal patrons, Isabella and Ferdinand, is the most pathetic thing he has read. Columbus claimed to have been very close to rediscovering the original Garden of Eden, in today's Venezuela, and it was his detailed description of the ships smoothly rising toward the sky, when the weather got milder as they drew closer to Paradise, that especially irritated Warren. He believed Columbus was bluffing to impress his patrons, and to secure new funds for one more expedition—this time to lead the Spanish ships all the way to Paradise.

I find it interesting that Warren broke with the idea that explorers could sail or travel somewhere to find the answer to the origins of man. Today, after excavations in Ethiopia, Tanzania, and South Africa, where the skeleton named Lucy was retrieved in 1974, the "out of Africa" theory is widely accepted. For Warren, it was a matter to be reflected on, pondered, and studied, rather than observed. According to him, the Greek poet Pindar had also understood this, when some 2,400 years before he had written:

> Neither by taking ship
> Neither by any travel on foot,
> To the Hyperborean Field
> Shalt thou find the wonderous way.

Warren's book opens with the sentence: "This book is not the work of a dreamer." He refers to a number of religious and historical sources in his attempt to prove that places such as Atlantis, the Garden of Eden, Mount Meru, and Hyperborea were all at the North Pole, and that the myths that underpin each of these areas have shared prehistoric and geographical roots.

In our time, the planet Mars has somewhat usurped the North Pole's position as the new frontier—a destination for explorers willing to risk their lives. In 2022 and 2023 I was engaged by the European Space Agency to advise on how the first European could reach the moon, and later Mars. Such an adventure has many similarities to a traditional expedition north. You sleep in a tiny space, it is dangerous outside, the food tastes the same, and the future is uncertain. Astronomers believe that the answer to the origins of life may be found on Mars, a planet where the temperature at the poles can drop to –243°F and the journey each way is at least 35 million miles. The idea of a manned expedition to Mars in the next ten or twenty years to explore, colonize, and find the answer to some of humankind's most pressing questions might seem as exotic to us now as an expedition to the North Pole did in the 1800s.

In his book, Warren refers to "scientific papers" that confirm that the climate at the North Pole was once warm, and that it was a fertile, green place all year round. He also cites a Swiss scientific study that claims that the origins of all flowers in the world lie at the North Pole. A third study concluded that the northern region was the original home of all animals.

The professor also points out in his book that in the first known religion in China some 4,000 years ago, the celestial North Pole was seen to be the center of heaven, where the gods live. Other religions in India, Egypt, Iran, Greece, and Japan had similar beliefs. There is a paradise on earth at the North Pole, Warren claims, and another directly above at the celestial North Pole. According to his reading of the Talmud, the two paradises are connected by a pillar;

in other words, the world's axis continues up vertically from the North Pole.

Warren believed that not only mountains but also trees were of significance to understanding the North Pole—many different cultures talk of a cosmic tree as a link between heaven and earth. In Germanic traditions the tree is called Irminsul, or "heavenly pillar." We are familiar with this idea from Old Norse mythology, where the ash tree Yggdrasil is called the Tree of the World. This tree stands at the center of the world. Its branches stretch up into the heavens and spread throughout the world.[47] Like Hyperborea and Mount Meru, no living being has seen any holy trees.

The North Pole as a State of Being

When I returned from the North Pole in 1990 and gave talks about the expedition, I concluded by saying that even though we now live in a world where so much has been discovered, I hoped that everyone would discover their own North Pole. I still like the thought of finding your own path, of traveling to your inner pole. And have since understood that this idea is in fact an ancient one.

Nine hundred years ago, the Iranian Sufi master Suhrawardi (*c.* 1154/55–91)[48] wrote about setting forth on a spiritual journey where the power of prayer, color, and the energy of light would help people overcome their inner darkness and lead to a heaven of pure immaterial light. In Sufism, worship takes the form of fixed rituals—*dhikr*, which can be translated as "remembrance"—often combined with breathing techniques and physical movements that create spiritual energy. I have tried it, breathing deeply in and out through the nose, dancing a sedate dance, and repeating the same words. It works—if you believe in it. There was a shift in my state of mind, though I did not reach the state of a Sufi master.[49]

Such a state will eventually lead to the light that Suhrawardi wrote of: a lost paradise where the first humans lived. The journey there, according to the French Professor in Islamic studies Henry Corbin,[50] was not to a geographic North Pole, but rather to a personal, inner pole.[51]

Corbin, who was also a philosopher, based his work on Suhrawardi's research on Zoroastrian Islamic wisdom from the time before the Prophet was born, as well as people's ability to travel to a North Pole beyond the North Pole, in their mind.

The Sufi master has been called the Master of Illumination, but his teachings on how ancient wisdom could be combined with Islam were controversial, and he eventually suffered a martyr's death in Aleppo. His contemporaries believed that the light comes from the east, where the sun rises, and while Suhrawardi did not disagree, he believed that the light of Sufism was something else, and far greater than any physical light. It could not be seen with the naked eye, and the source was a perfect nature where everything was connected, a place at the center of the world and its greatest secret. Professor Corbin's interpretation of Suhrawardi is that this wisdom comes from the cosmic north, which lies beyond the most northerly point on earth.

Suhrawardi and Corbin compare seeking this light with climbing the sacred mythic mountain Qâf, which in Arabic and Iranian cultures is the highest mountain in the world, at the farthest point on earth, most often in the north. In Arabic, *noor* means "light" or "sublime light." Legend has it that there is an emerald rock at the top of Mount Qâf that creates its own light. Anyone who has seen the Northern Lights knows they are most often innumerable tints and variations of emerald green.

Even though it could be fatal for Sufis to promote ideas that conflicted with accepted doctrine, Suhrawardi was not alone. Another Sufi master, Najmuddin Kubra (1145–1220/21), like Suhrawardi, also placed great importance on the immaterial light and how a person of faith can experience that light through various levels of spiritual practice. For Kubra, it was an inner journey north and upward. First you experience darkness, then you realize it is a black sky; the next stage is red, then white and a white cloud, then dark blue. The colors shift, depending on where you find yourself. In the Koran it says: "Light upon light" and no shadow.[52] Kubra emphasizes that the colors mentioned are of a lower spiritual level. Emerald green reflects the highest level the soul can attain, it is the color of the heart and of life, and green lasts longer than any other color.

Kubra maintained that green was the color at the top of Mount Qâf and often the color of the celestial North Pole. And while the devil lives in the dark, angels live at the highest point, in the light you are striving to attain.[53]

The journey through your consciousness must be taken alone, according to Suhrawardi, and it is when you feel the discernible and pure green light that you discover your inner pole—your true self.

The people who lived closer to the North Pole did not share these ideas of the imaginary North Pole. They knew the area better than the Indians, Egyptians, and Sufis, and were used to seeing the Northern Lights, and so did not share their philosophies of a positive, spiritual journey north. The myth of Hyperborea must also have been peculiar to those who lived in the north, if they knew it at all. The northern dwellers were often pithier in their description of the world in the north. For example, in the Finnish national epic poem *Kalevala*, it is said:

> . . . dark Northland,
> the man-eating, the
> fellow-drowning place.[54]

EIGHTEEN YEARS AFTER my parents gave me the globe, in 1988, I went on my first little expedition to the Arctic, to prepare for our expedition to the North Pole. During those days and nights in Svalbard, I understood why the Vikings called the Arctic "Home of the Fog." Northern dwellers must have sailed close to the pack ice more than 1,000 years ago and experienced the climate and the freezing fog themselves.

On that first expedition, I realized that the fog was not the hardest thing about walking to the North Pole. It was not the great distance, not the 265 pounds on each sled, not the drifting ice, nor starving polar bears. It was simply getting up each morning. When it is between −60 and −40°F outside the tent, it is all too tempting to stay in your sleeping bag for just another five minutes, rather than crawl out shivering and freeze like hell. That has been—and remains—the greatest challenge since polar exploration began.

2

A REVOLUTION IN PERCEPTION

A mid-nineteenth-century lithograph of Henry Hudson, set adrift with his son, John, and seven crew members after a mutiny on his ship Discovery *on June 11, 1611.*

Father of Extreme Tourism

It is impossible to know when people started to feel the pull of adventure, and chose to climb mountains, venture into forests and over the ice, when they were first captivated by nature's wild beauty. The tradition of going outdoors to enjoy nature, and not for hunting or gathering, is generally thought to have developed in the late Middle Ages, but I find it hard to believe that for thousands of years people were untouched by nature or the wish to explore when out collecting food and wood. There is something deeply human about delighting in a sunrise, in hearing the birds sing and feeling, the onrush of running water in a nearby stream.

ON APRIL 26, 1336, the poet Francesco Petrarch climbed Mont Ventoux, a mountain in Provence, France, with a relatively easy ascent, 6,266 feet above sea level. He was accompanied by his brother and two servants and had chosen a dangerous and all-consuming goal. He said he wanted to climb the mountain so he could experience getting to the top. In that moment, he became the father of extreme tourism.

As soon as he descended, he wrote "Ascent of Mont Ventoux," and there is a sentence in the letter that both surprises and thrills me: "My only motive was the wish to see what so great an elevation had to offer."[1] This, and the fact that he had voluntarily struggled to the top, may perhaps be the first written account of choosing to be in nature for personal satisfaction and enrichment, a time for inward reflection. Modern polar explorers are very much a part of this tradition.

For me, Petrarch's letter is about adventure. He chooses to leave the comfort of his home and relinquish control over his fate and surroundings—to dream, to be hungry, to be full of doubt. In my experience, home is not somewhere we go without food and drink or feel unsafe. Yet when these are lacking, life seems somehow richer. When it is no longer a given that these primary needs will be met, you become acutely aware of them.

He thinks about the food he is going to eat as he descends in the moonlight and knows that it will taste all the better for having gone without it. It is as though he realizes that he has all he needs in life, but everything can be lost, and so becomes more valuable.

These days, more or less anyone who knows what to expect could climb Mont Ventoux, but when Petrarch speaks of his first ascent, he mentions experiences and emotions that most explorers will recognize. Curiosity and wonder, a clear goal, unknown territory, the desire to be first, or nearly first, the relief of leaving one's worries behind, and the nagging question of who to take with you. Petrarch tries to find shortcuts on his way up the mountain, but these prove to be detours. When his party then meets a local man who strongly advises them not to go any farther, warning that the only thing they will gain by carrying on is regret and exhaustion, it only spurs them on. Yet in the end, after the expedition, he feels empty. Most polar explorers will be familiar with this feeling of emptiness; you have been consumed by a dream day and night, an obsessive love for a distant goal, and then suddenly it is no longer there.

Petrarch covered great distances on foot in Belgium, Italy, and France. He wrote love poems, sonnets, and letters, and talked to people—his writing clearly reflects the fact that when he moved physically, he was moved emotionally. He also took the opportunity, while walking, to buy classical literature along the way, and his collection grew into one of the most extensive libraries of the fourteenth century.

Petrarch has also been called the father of humanism. He emphasized the importance of classical knowledge and promoted universal education, inspired by classical Greek and Roman literature, art, and culture. He believed that knowledge was gained through personal choices and experience, and through conversing with people, not with God, thus breaking away from the Western tradition in which he had grown up. For Petrarch, it was crucial to place people, rather than God, at the center of life.

To climb a mountain, as described by Petrarch, is also part of the Christian mystical tradition of meeting God in elevated places, in or close to heaven. But unlike Dante in *The Divine Comedy* or Francis of Assisi at Monte Penna, God did not talk to Petrarch. He did not experience light, peace, and truth.

When he reached the peak, Petrarch was overwhelmed by the view of a landscape he knew so well from the ground, but something was missing. He took out his copy of St. Augustine of Hippo's *Confessions*, a book he always carried with him whenever he went walking, opened it at random, and read: "Men go about to wonder at the heights of the mountains, and the mighty waves of the sea, and the wide sweep of rivers, and the circuit of the ocean, and the revolution of the stars, but themselves they consider not." He closed the book in agitation and deep existential doubt. The text made Petrarch realize that he was more concerned with what is around him than within him. The only real revelation he had at the top of the mountain is how little he actually understands.

Petrarch's doubt is reminiscent of Socrates's and his statement that the only thing he knows is that he knows nothing. This doubt is fueled by desires, mistakes, passions, and shifting sensory impressions.[2] As a polar explorer and mountaineer, I recognize that feeling of doubt, which stems from something other than mere uncertainty. When you are out on the ice or climbing a mountain, you can experience moments that seem to last an eternity, moments when your mind is blank, you simply experience. The past and the future mean nothing; you are fully present in your own life in that moment. Only those who can live in the moment can be sure of eternal life.

Petrarch and anyone else who likes to hike does so for more than one reason. And yet his one motive for climbing Mont Ventoux—to simply see the view—strikes a chord and preempts George Mallory's famous answer to the question why. "Because it's there." For me, both show how little there is that we *must* do in life. No one *must* climb Mont Ventoux or go to the North Pole. There is much that we should do, *can* do, but little that we *must* do. If you choose to go to the North Pole, things might turn out well; if you sit at home in your chair, all will probably be well. But things that come too easily are soon taken for granted. There has to be a price, some form of discomfort—the cold, wind, thirst, and steep slopes. Satisfaction comes from continuing to move laboriously forward in the right direction, no matter what. Not knowing if you will reach your goal. That is when life feels real. Time expands.

IN THE TIME since Petrarch wrote *Ascent of Mont Ventoux*, there have been those who questioned if he actually climbed the mountain, and those who have accused him of bragging about the final stretch, in the same way that some people question polar explorers who claim to have reached the North Pole. I see no reason not to believe that he climbed the mountain and did not just make up the feeling of emptiness.

Francesco Petrarch was not only a poet, but also a wayfarer. He gathered knowledge by taking one step at a time, his account is realistic, and he thought of himself as a wanderer to the very end of his life. One of the last things he wrote, as he tidied his things, before his life ebbed away, was: "I am arranging my belongings in little bundles, as wanderers are wont to do."

A RED THREAD runs from Aristotle's belief that we should wonder at our world and experience it directly with our senses, to Pytheas's voyage north and Petrarch's letters. The thread continues on into the Renaissance, an era that spawned European ambitions to conquer the world and subjugate nature.

The English author Roland Huntford, who has written a number of books about polar explorers, maintains that expeditions in the Renaissance to discover, trade, convert, and colonize created the foundations for an attitude that shaped our Western culture and laid the basis for polar expeditions.

According to Huntford, the polar expeditions that followed these voyages of exploration are also a Western phenomenon. The men who, "with varying success, pitted themselves against the hostile polar environment were the heirs of the Renaissance, the heart of which was the promotion of the individual and the discovery of the world. They represent our kind of civilization."[3] Regardless of whether you would call it our civilization, I think he's right. The demand for maps and globes in Europe grew in the wake of these long journeys. European cartographers started to draw and define the world based on the land, islands, and oceans that explorers described on their return, and what they believed lay beyond what they had seen.

Gerardus Mercator—The World Is Redrawn

In my twelve years at school, wall maps printed on thick paper were a part of everyday life in the classroom. When the colorful, seven- to ten-foot maps were not in use, they were rolled up like blinds under the ceiling at the front of the classroom, just above the blackboard. All the teacher had to do was reach up and grab the string, then she could reveal a world to us: Europe, Scandinavia, or Norway, as she talked about religion, wars, politics, migration, and exploration.

At the end of the 1990s, I heard that schools in Norway were getting rid of these fantastic maps. So I decided to get myself a world map before they disappeared and rang my old primary school, but I was too late—the maps had already been thrown out.

Perhaps it was a sign of the times. Someone had decided that my daughters' generation should not have the world around them as a starting point, or measure of scale, when learning geography. More often than not, digital maps put the viewer at the center. An on-screen map does not show where you could be tomorrow, or in your dreams, since more often than not it starts by showing where *you* are now, "Marked by a blue dot that offers no resistance, but happily follows you wherever you go," as the teacher Andreas Brekke wrote in the Norwegian newspaper *VG*. When he returned to the classroom in the 2000s, he discovered that the maps that had allowed him to experience the world and dream about the future were gone.[4] The Google maps that my daughters are growing up with were not made by cartographers, but by engineers and algorithms, and seem dull and flat. And because they are on-screen, you are dependent on electricity and a good signal, and they are small—I feel they exclude so much of the world.

The dimensions of the world map were a mystery at school. Why was Antarctica so big that it filled the bottom of the map and why was the white island Greenland, in the far north, bigger than Africa? Even though Africa is bigger than Greenland? Had we learned about the geographer, cosmographer, philosopher, and mathematician Gerardus Mercator (1512–94), we might have understood this lack of accuracy better.

Mercator has been called the father of modern cartography. He was ten years old when Magellan's ship *Victoria* completed the first circumnavigation of the world in 1522. Naturally enough, the event generated a surge of new interest in maps. Mercator's father died five years later and the fifteen-year-old was taken in by an uncle, who hoped that his nephew would become a priest, just like him. But Mercator began to question the Catholic Church's worldview. Instead, he studied mathematics, astronomy, philosophy, and—fortunately—geography.

When he became a cartographer, Mercator believed that a world map should be vast, in order to give a vision of the world, which was unusual at the time—just as it is once again now. Three copies of Mercator's world map still exist, and the best preserved is kept in the University Library Basel. It is 79.5 inches wide and 53.1 inches tall. No printmaker had a press big enough to print the whole map, so it is composed from eighteen separate sheets. Mercator not only drew the map, but also added facts and information, a total of 5,000 words, all in Latin.

WHEN I WAS growing up, we held physical maps in our hands when we navigated on land or sea. I wanted to feel Mercator's map, to hold it—to experience its true scale and study the continents, countries, and oceans as I had learned to do many years ago. To smell the paper and perhaps even discover some details that were lost on-screen.

In summer 2022, I went to Basel to see the map.[5] I hoped I would find the answer to something I had been wondering about. When I arrived, I walked from the train station straight to the library. Then, at the entrance, I turned on my heel and walked around the building before going in. I like to do that, when expectations are running high, to take a few minutes more to remind myself of what I might experience. By walking around the building and then through the Botanic Gardens that happened to be next door, I circled the map, put it in the center.

As I walked up the steps and into the building with librarian Dr. Noah Regenass, I mentioned how tickled I had been to discover that the library and the Botanic Gardens were neighbors. Regenass stopped, turned, and paraphrased Cicero with a wry smile: "If you have a garden and a library, you have everything you need."

The map is kept in three gray storage cassettes. Each case holds six sheets that are glued together. When I got out a pen to take notes, Regenass gave me a friendly look, and without a word, handed me a pencil. Once I had put on some latex gloves, Regenass asked me to help him unfold the map. I was almost afraid to touch it. When I opened the map, I pressed my nose briefly to the paper and inhaled deeply to see what a world map from 1569 smelled like. Nothing. Not even the faintest whiff after having been locked up in an archive. The map was as devoid of smell as the air at the North Pole, where all particles are frozen to ice.

I asked Regenass why, and he explained that light, dust, and moisture can spoil maps, and it is moisture that makes books and maps smell. The map is stored in such a way that there is no risk of damage and only taken out three times a year to check its condition.

The first thing that struck me when I looked at the map was how detailed and well defined the area around the Mediterranean was: North Africa and the Middle East looked more or less as they do on modern maps; Italy was shaped like a leg kicking a football; southern Europe, the British Isles, and Ireland were all recognizable. In a biography of Mercator, the British author Andrew Taylor comments that all the towns and cities marked in Europe, Asia, and Africa demonstrate confidence in "a landscape that has been tamed."[6] The white areas, which had not yet been explored by the Europeans, reflect a belief that European culture will flourish there too. For them, it was only a matter of time.

Parts of North America and the whole of India are both given the name India on the map, and Australia was not included. Magellan had found and sailed through his strait by then, but he had not yet located Cape Horn, so South America is attached to Antarctica. As Europeans had not explored the interior of South America, three scenes are drawn near Patagones. A naked person chopping up another person with a huge knife; a tree with human parts hanging to dry; and a final drawing of two naked people grilling meat over a fire. People living in areas that were still to be Christianized, or "civilized," have, throughout history, been called barbarians, and their countries portrayed on maps as a hell teeming with monsters, Amazons, and cannibals. One reason for these depictions was to create the illusion of superiority, and for Europeans to go there and proselytize.

Yet Mercator's map had a weakness that was also a strength. The longitudes were shown as straight and at a fixed distance from each other, unlike those in the real world, which means that they do not meet at the North Pole, but continue on into eternity at the same distance as at the Equator. In short, the world is expanded at the poles and shrunk in the middle. Mercator's projection—his mathematical method for transforming the globe, or smaller parts of it, into a map—is used as the basis for nearly all maps in modern times. The modern maps are of course much better, but this fundamental flaw remains the same. I still remember being irritated by how enormous Norway, the Soviet Union, and North America were on the world map in our classroom. The Arctic and Antarctica stretched along the top and bottom edges of the map, taking no account of the curve of the earth.

The cartographer was of course aware of this weakness and in the bottom left-hand corner has drawn a small map of the North Pole, where the longitudes meet. However, he still believed that the exact position of the North Pole on the world map was not important, as no one would travel there anyway. For several hundred years, hardly anyone disagreed.

Mercator was more concerned about the usability of the map. One strength is that his projection is well suited for navigating great distances at sea, as long as it is not too far north or south. Mercator drew lines—rhumb lines—on the map that show the most efficient direction to sail from one point to another.

Next to the map of the North Pole, Mercator writes that he has used two sources for this part of the world map, both of whom were familiar with the area. The most important source was a mathematician and Franciscan monk who Mercator believed had written the mythical *Inventio Fortunata*—"A Fortunate Discovery"—which contained a number of facts about the area around the North Pole. The book, however, had already been lost.

According to the text on Mercator's map, the monk traveled north beyond the Hebrides and Iceland at some point in the early 1360s and then carried on with the help of "arte magica" to explore and map the far northern areas.

No one knows what was written about the North Pole in *Inventio Fortunata*, but we do know what Mercator was told: that the North Pole was an island, Rupes Nigra, a black magnetic mountain that rose up into the clouds and heaven.

To imagine that the North Pole was a dark mountain of iron was logical enough. People had already been using lodestones—pieces of the brownish-black mineral magnetite, a naturally occurring magnet—as a compass for thousands of years. A lodestone attracts other magnetic objects and will point north when hanging freely in the air.

Fortunately, the librarian Regenass had a magnifying glass to view the fine script. I held it up to my right eye and leaned over the North Pole. The polar point was encircled by a kind of whirlpool and four islands, with four rivers running between them. On the map, it says: "These four arms of the sea were drawn toward the abyss with such violence that no wind is strong enough to bring vessels back again once they have entered."[7] Water and ships are swallowed into the depths around the North Pole, and new water runs down from the mountain, as from a cistern. According to the text, the wind at the North Pole is weak and not strong enough to turn a windmill.

The other source to which Mercator refers was a book by the well-traveled Welsh priest and historian Gerallt Gymro, or Giraldus Cambrensis (1146–1223) as Mercator called him. Cambrensis was in his day described as "one of the most learned men of a learned age."[8] It is unlikely that Cambrensis traveled north of England, but he confirms Mercator's theories about the North Pole and is quoted on the map. Beyond the Hebrides and Iceland "toward the north there is a monstrous gulf in the sea . . . should it happen that a vessel pass there, it is seized and drawn away with such powerful violence of the waves that this hungry force immediately swallows it up never to appear again."

The Vikings also told similar stories about undertows, or "the sea swallower," in the far north. These were so wide and deep, and the whirlpools at the center so powerful, that any ship that managed to get that far north would never return.

ON THE MAP, to the east of northern Norway (which is called Lapia and Lapp en land) four nearby islands are described. One of them is inhabited by people who are four feet tall, *Pygmei*—the same people as those called *skrællinger* who lived on Greenland. This is probably derived from the Old Norse sagas that describe the indigenous people of Greenland as *skrællinger*, a derogatory term meaning "small barbarians." The Vikings used the same name for the people they met when they rediscovered America.

It was not easy to tear myself away from the map. To the east of Lapia—an area at the opposite end of Europe from Herodotus's Halicarnassus and Athens—I found a reference to Hyperborea and three words that, according to Regenass, referred to the mountain range that separated the known world from Hyperborea, Camenon pyas mon.[9] Although there was some disagreement in antiquity about where Hyperborea was, most people agreed with Herodotus that it lay north of an almost unknown mountain range, which is included on the map, following the northern edge of the world.

Mercator has drawn Poseidon, god of the sea, in the Pacific Ocean, and there are other sea monsters scattered throughout the oceans and seas. On older maps, I have seen monsters that are drawn to look threatening, as a warning to anyone who might be considering a long voyage. In medieval Europe, it was unusual to think of exploration and research as separate from the Church, which is one of the reasons why gods, myths, dragons, and sea monsters were included on maps.

BECAUSE ARISTOTLE HAD written that the area around the Equator was uninhabitable, Europeans believed this for the next 1,700 years. Aristotle divided the world into five climate zones, determined by their distance from the Equator. It was too cold and there was too much snow and ice to the north of the Polar Circle for anyone to live there. And it was impossible to survive at the Equator, because it was too hot.

The Arabic cartographer, chronicler, and explorer Ibn Hawqal (*d.* 988) spent thirty years of his life exploring Africa and Asia. When he traveled south through Africa in the 900s, he saw for himself that Aristotle had been wrong. He crossed the Equator and observed that it was a good place to live.

It was not until the first European explorers crossed the Equator toward the end of the fifteenth century that they also realized that Aristotle had been mistaken. They discovered animals, countries, and people who were not mentioned in the Bible, at latitudes that should be uninhabitable, according to the great philosopher's logic. The need for new sources of information and new maps became increasingly clear.

JOSEPH CONRAD (1857–1924) wrote in an article in *National Geographic* magazine in 1924, that cartographers, like other scientists, found the truth through "a long series of errors."[10]

Conrad divided the history of cartography into three eras. The first he called *Geography fabulous*. It was a time when cartography was colored by "extravagant speculation which had nothing to do with the pursuit of truth, but has given us a curious glimpse of the medieval mind." Wild animals and monsters were included as a part of reality on maps. This era had waned after Europeans sailed to America.

The next era was *Geography militant*, from around the time of the early exploitation of Africa, and includes polar explorers "whose aims were certainly as pure as the air of those high latitudes where not a few of them laid down their lives for the advancement of geography." This is eloquently put, if not entirely true. As we shall see, polar history is full of explorers with less than noble motives.

Conrad felt that he was part of the third era, *Geography triumphant*, the modern world. The age of great heroes was over, and new explorers followed well-trodden paths, like an ordinary traveler.[11]

IT IS FASCINATING to remember that two of history's greatest cartographers, Mercator and Ptolemy, who lived approximately 1,400 years apart, hardly ever traveled. One of Mercator's longest ventures was to sail up the river Main to Frankfurt in 1554, to visit the fair for books and maps.[12] He sat indoors and imagined how the world must look. Like Ptolemy, he was one of the great scientists of his day, and his library contained more than a thousand maps and books. Both Mercator and Ptolemy were interested in mapping the stars, but as the telescope would not be invented until 1608, their sources were limited. They also believed in Aristotle's idea that the world had to be experienced to be understood, but left the experiencing to others.

Mercator corresponded with some of the world's leading mathematicians and cartographers, from Portugal in the west to England in the north and India in the east. He studied the accounts of explorers such as Vasco da Gama, Columbus, Magellan, and Marco Polo. His son, Rumold, lived in London, where he worked for a bookseller and so was able to get accounts and hand-drawn maps from English voyages north.[13] As King Manuel of Portugal had introduced a death penalty for anyone who divulged the geographical details of da Gama's sea route to India, and the Spanish king and queen, Isabella and Ferdinand, did what they could to keep the areas discovered by Columbus secret, other countries employed spies to get the necessary information for their cartographers. However, this got easier over time, and Mercator was constantly finding new sources. Geographical discoveries were being made just as the maps were being drawn, and he constantly had to update his maps and globes.

PTOLEMY'S TEACHINGS LIVED on in the Islamic world; key works on astronomy were translated from Ancient Greek into Arabic in the ninth century and the centuries that followed. However, Ptolemy's *Geographia* was not translated into Latin until 1406. Ptolemy had been neglected for more than a thousand years until he was rediscovered in the Middle Ages, and was then revered by sixteenth-century cartographers. As the printing press had been invented by then, copies of his books (the originals had been lost) were produced and distributed. He was so popular in Mercator's day that you could buy etchings of his portrait to hang on the wall, and cartographers often included his portrait in a corner of their maps. Many of the 8,000 coordinates that he had given were used for new maps. Several of the coordinates were imprecise, and so the errors lived on.

IN 1494, SPAIN and Portugal signed the Tordesillas Treaty, which divided the newly discovered—non-Christian—world in two. A vertical line was drawn from the North Pole to the South Pole down through the Atlantic Ocean. Everything to the east of the line—for example, Africa—came under Portuguese control, and everything to the west, with the exception of Brazil, fell under Spanish control. The line had originally been proposed in a papal bull the year before, but Portugal and Spain were not satisfied with the pope's decree and continued to negotiate with each other. They saw the world as a hunting ground that was theirs to divide and conquer—and assumed that the treaty would last forever. Most of the new world that was to be carved up was completely unknown to both empires. Not only did they split discovered and undiscovered islands and mainlands, but they also divided the trade routes over the Atlantic, and subsequently over the Pacific and Indian Oceans. Thus a substantial share of world trade was monopolized by the two empires.

Problematically, the treaty was signed before the two parties had agreed on the size of the world. If Columbus had been to India, as he was erroneously convinced he had, the distance from Europe to India was no greater than from Europe to America. This persistent belief that the world was smaller than it actually was meant that these maps were of little practical use. Yet they continued to work together to maintain their monopoly and prevent other nations from sailing in their areas.

England and then the Netherlands soon turned their focus north in the hope that they could find a new route to the Pacific Ocean, crossing above the north end of America. No one knew where the continent ended. The idea of sailing over the North Pole, or finding a northwest or northeast passage to compete with Portugal and Spain without interfering with their trade routes, was born.

PTOLEMY AND MERCATOR had the same mission in life, which seems to be impossible: to create the perfect map.

The world's most influential producer of maps today is no longer a cartographer, but Google. As Mercator's projection continues to be used, it is still not possible to position the North Pole on the world map. Google maps, of course, have far more facts, priorities, and details than the maps we had at school, but contain many of the same errors. Antarctica is given too much space in the south in relation to the continent's size, and the map stops just north of Canada, Alaska, Svalbard, Greenland, and Russia, as Mercator's did in 1569. Google maps remind me of maps that were made before Ptolemy, when the North Pole was not included at all. Back then, cartographers knew nothing about the North Pole, other than its possible existence, so it was simply omitted. Today, however, the North Pole is not included because Google engineers, like Mercator, are still not able to make a map of the world where the longitudes meet at the top.

When I study the Google world map, I am reminded of the question that all great cartographers have pondered, which even Google does not have the means to convey or answer: What are the outer limits of the world?[14]

The Idea of Sailing to the North Pole Is Born

I believe that the English merchant and cartographer Robert Thorne was the first to have the defined goal of reaching the North Pole and to make plans for a North Pole expedition. Thorne approached the court of King Henry VIII in the 1520s, and finally wrote to the king himself, saying that England could not simply sit by and accept Portugal and Spain's role as the dominant players in world trade.

The American continent was a barrier between England and the Pacific Ocean, and no one knew where it began or ended in the north or south. Thorne proposed an expedition north to find sea routes to India, China, and the Spice Islands (now known as Indonesia), via the North Pole, which he presumed was surrounded by open waters in summer. He was driven by the scent of spice and the lure of gold and new countries. A common belief at the time was that not only spices but also gold was abundant on the other side of the world thanks to the warmth from the sun. And a northern route would be several thousand nautical miles shorter than the routes controlled by Spain and Portugal, and the time saved would obviously offer a huge advantage in terms of competition.

Thorne highlighted one of England's weaknesses. Portugal and Spain had become two of the most powerful nations in the world in the course of the fifteenth and early sixteenth centuries. Within a matter of decades, Portugal had gone from being one of the poorest countries in Europe to one of the richest in the world. Wealth was extracted in the wake of the great explorers. They crossed the Atlantic to Brazil and sailed south along the west coast of Africa. It was the Portuguese who first rounded the Southern Cape and sailed on to the Indian Ocean. They established trading stations and colonies in India, Brazil, and Africa, and on the islands of Cape Verde, Madeira, and the Azores.

Explorers crossed the Atlantic on behalf of the Spanish crown, and colonized what is now Mexico and large swaths of South America and the Caribbean. This resulted in great wealth not only for both nations, but also for investors and enterprising merchants. The two countries had effectively globalized the world without England, and Thorne and many others watched in consternation as England became more isolated.

There are two versions of what happened with Thorne's idea. One, which seems to be more accepted today, is that Thorne bought what he believed was the perfect ship, *Saviour*. She was one of the biggest ships ever built in southwest England.[15] A captain was hired and the preparations to make *Saviour* fit for purpose were started, but then Thorne died suddenly. He was only forty years old. Without his drive and money the project lost momentum and *Saviour* did not sail north. King Henry VIII never received the letter and never got to hear about Thorne's ambition.

The other version is more exciting. The king appreciated Thorne's idea and plan, and wanted not one but two ships to sail to the North Pole. The British often preferred two or more ships, as they thought that wisest in case one got wrecked. A sensible thought, perhaps, but it also doubled the risk of a fatal accident.

The historian Richard Hakluyt (1552–1616) wrote about the expedition both in 1589 and in 1600. To appreciate these works, it is an advantage to read the original texts in gothic script. Thorne was, according to Hakluyt, a "notable and ornament member for his country, as well for his learning," but he preferred not to participate himself. On board there were, according to the historian, "cunning men to seek strange regions."[16] The expedition set out on May 20, 1527. While crossing the Atlantic Ocean, the ships sailed too far northwest and found themselves in "a dangerous gulphe" between Newfoundland and Greenland, possibly the Gulf of St. Lawrence. And there they ran into Portuguese and French fishermen.

One of the ships sank and any accounts that were written about the expedition, either en route or after, have been lost—as Hakluyt writes—"by reason of the great negligence of the writers of those times, who should have used more care in preserving the memoires of the worthy actes of our nation." There is, however, little reason to believe that the expedition returned with full coffers or information about new trade routes, and so was deemed a failure.

For me it is not so important which of the two versions is correct. However, the uncertainty as to what may have happened almost 500 years ago, when the North Pole was the goal, foreshadowed a question that has dominated polar history: What actually happened along the way?

TWENTY-SIX YEARS PASSED before another attempt was made. In 1553, an English expedition set off to find the Northeast Passage by following the north coast of Russia, rather than going over the North Pole. The expedition was financed by a newly registered limited company with an unusually descriptive name: Mystery and Company of Merchant Adventurers for the Discovery of Regions, Dominions, Islands, and Places Unknown. The company later shortened its name to the Muscovy Company. The expedition's primary goal was to make money for its investors. It was led by the Englishman Sir Hugh Willoughby, who was chosen "both by reason of his goodly personage, as also for his singular skill in the services of war." He received his knighthood after having distinguished himself leading English troops in the burning of Edinburgh and other cities in fighting against the Scots in May 1544.

The three ships—*Bona Esperanza*, *Edward Bonaventure*, and *Bona Confidentia*—were towed down the Thames on May 10, as the crowds cheered from the banks. The ships' names reflected the general optimism—hope, expectation, good luck, and daring—as the naming often did on polar expeditions. The ships were equipped to deal with extreme conditions, and to withstand the cold and ice. There was obviously great seriousness about the expedition's chances of success as one of the ships even had its keel coated in lead to avoid damage from shipworms when they got to the Indian Ocean.

Put simply, the ships' progress north was disastrous. The intention was to reach northern Norway within a few weeks, but they did not get there until July 14. The captain of *Edward Bonaventure* is said to have named the most northerly point of Norway North Cape. They must have been extremely confused when the compass needle no longer pointed to the North Pole and North Star, but instead pointed west to the magnetic North Pole, which had not yet been discovered. They would therefore not know that the deviation—that is to say, the difference between the direction of the North Pole and that of the magnetic North Pole—was considerable. Another challenge was that they were using early versions of Mercator's map and globe. Mercator knew hardly anything about the geography of that part of the world, so the map was almost no use at all. In fact, it would be harder to navigate with the map than without it.

Bona Esperanza and *Bona Confidentia* were stranded on the coast of the Kola Peninsula in Russia, after they passed the southwest tip of Novaya Zemlya, the island just north of Russia. Everyone on board died, including Willoughby. For a long time, it was thought that they froze to death, as would be natural in the Arctic—this was in the time of what was later called the Little Ice Age—but the cause of death was more likely something else.

In the spring of 1554, the year after the ships were separated from *Edward Bonaventure*, they were discovered by Russian fishermen. The fishermen called out and shouted, without getting a response. So they boarded the vessels to look for the crew. All the doors and hatches were closed, and they had to force entry. They found the men in the salon and cabins, some sitting as slumped sculptures at the table where food had been served, others lying on their beds. Even the dogs on board had died in natural positions. The men, dogs, food, and water were frozen solid. Willoughby was reported to have been sitting at the table with his diary close to him. The last time he wrote anything was in January 1554, when he witnessed a will.[17]

It seems likely that the cause of death was in fact carbon monoxide poisoning. To keep themselves warm through the winter, the crew had insulated the ship so well that no fresh air could get in, and they were burning coal. They had even insulated the chimney.[18] They would never have experienced cold like it, and probably had no idea how easy it is to get carbon monoxide poisoning. When it is cold, it is only too tempting to keep out the fresh air in favor of the warmth that comes from a burner or stove. The danger, of course, is that carbon monoxide poisoning is a silent assassin. If you are asleep, you could die without noticing, and if you are awake, you might cough, get a headache, but think that it is something else entirely. Most people I know who have slept in a tent or drafty cabin had been tempted to shut out the fresh air and keep in the warmth. The trick is not to do it too well.

The third ship, *Edward Bonaventure*, managed to get to current-day Archangel, in Russia, where they cast anchor. The captain traveled to Moscow, 631 miles away, to meet the tsar, Ivan the Terrible, and was surprised to discover when he got there that Moscow was bigger than London. Ivan was Russia's First Tsar and God's Representative on Earth and he was keen to establish new trade routes. The Swedish empire and Polish–Lithuanian Commonwealth were a barrier to Russian trade with Europe. In addition, large parts of northern Europe were controlled by the Hanseatic League. Ivan and the captain signed an agreement that granted the Muscovy Company a monopoly on trade between England and Russia, which would last until 1698 with only a few interruptions.

The primary goal of the expedition—to make money—was achieved. Despite the death of so many men, it was seen to be a huge success in England. This inspired entrepreneurs.

THERE IS AN old English saying that "every pirate wants to be an admiral." England's great polar hero in the years that followed, Martin Frobisher, was a perfect example. He was not only an explorer, but also a pirate. Sometimes he plundered ships on his own behalf, and other times on behalf of Queen Elizabeth I. The queen was so pleased with his discoveries and the loot he returned with that she appointed him vice admiral.

Frobisher led three expeditions to find the Northwest Passage and discovered new territories, but it was never his plan to sail farther north and reach the North Pole. In 1578, the queen started to draw up a plan to include the new territories discovered by Frobisher in her empire. Frobisher had already claimed sovereignty over them in her name, and the queen had named the area Meta Incognita—the "Unknown Coast." A comprehensive document was compiled, with a series of maps and a case for why a large part of the unknown world, including "Groenland and all northern isles compassing . . . even until the North Pole" should belong to England.[19] Other nations did not support England's claim, of course, but the notion that England and the North Pole belonged together was formed.[20]

The Dutch—Heroic and Suffering Polar Explorers

In the time after Mercator's world map was published in 1569, the dream of traveling north started to flourish in Europe.

The idea of a warm, sunny, habitable area around the North Pole, which Herodotus had written about 2,000 years earlier, continued to thrive in western Europe. No one knew what exactly surrounded the North Pole, and many still believed it was a mainland and mountain. Anywhere that allegedly had sun all day and night for six months must be warm.

The theory was that the Gulf Stream carried on north, and it was said that Russian whalers had experienced open sea beyond Svalbard. That might well have been true. It was generally known that the climate was colder in the north, but there might still be a temperate zone north of the cold north wind, surrounded by open sea and bathed in sunshine for half the year.

The great Dutch polar hero, cartographer, and explorer Willem Barentsz (*c.* 1550–97) sailed to the Arctic in 1594, 1595, and 1596. The purpose of each expedition was to find a route through the Northeast Passage and map it, so that Dutch ships could sail to the other side of the world.

The Dutch Republic was a young nation, established in 1581, and its leaders felt almost obliged to send expeditions north. The Spanish king and the Pope believed that the Dutch Calvinists were heretics and did not deserve to live, thus sparking the Eighty Years War, or the Dutch War of Independence as it is known in the Netherlands. For the Dutch, it became dangerous to travel through continental Europe. All ships that sailed in waters governed by the Tordesillas Treaty of 1494 risked being captured by Spanish and Portuguese ships, and the crews kidnapped or killed.

Traveling via the north seemed a more secure and reasonable alternative.

Even though the new Dutch Republic did not have much land, it was one of the world's most powerful seafaring nations, with excellent shipbuilders. Money and prestige were the primary motives for public and private investment in the three expeditions in which Barentsz took part.

The plan for the first two expeditions was to sail up around the most northerly point of Norway and then head east. Ice to the east of Norway stopped both voyages, and the investors lost all their money. The expeditions never got farther than the western end of the Northeast Passage, however, and the island Novaya Zemlya, which the

Dutch called Nova Zembla, in the northwest of the Russian Arctic. Novaya Zemlya means "New Land" in Russian.

A new route was then planned over the North Pole. The idea was to sail so far north that they would get past the ice, which was presumed to be no more than a belt, and sail on north over an ice-free Arctic Ocean. Once past the North Pole, the ships would then carry on south to the Pacific Ocean and China. They had fine woolen clothes with them, so they were ready to dress up and make a good impression when they arrived.

The whole enterprise was dependent on them finding a strait between Asia and America—a strait that was named the Bering Strait in the 1700s. The heart-shaped map drawn in 1507 by Martin Waldseemüller (1470–1520) was the first to include it. The map is also the first to use the name "America" for the Americas, in the area close to what is now Brazil. America had previously been called terra incognita. The world map covers an area twice as large as Ptolemy's world map.

To this day, no one knows how cartographers could have possibly known that there was a strait running from the Arctic Ocean to the Pacific Ocean. Where Waldseemüller found this out remains a mystery. The answer may be that people in prehistoric times traveled farther by sea and land than is generally believed. Thor Heyerdahl, my childhood hero, tried to prove several hypotheses about migration with his ocean voyages. He met a lot of resistance. Yet one of his fundamental ideas, that we underestimate our ancestors, seems profoundly true—given their sense of adventure, the boats they sailed, and the voyages they made. The Bering Strait may have been "discovered" several times in the course of history, and information about the passage passed down haphazardly through the centuries until it reached Waldseemüller.

THE APPEAL OF sailing north to find a new route to the Pacific Ocean and China, rather than sailing south to where the Catholics ruled Europe, emerged around the same time in Protestant England and the Netherlands. What was new about Barentsz's third voyage was that the expedition freed itself from the usual practice of sailing close to land, for the safety of the ship and crew. They sacrificed some of this safety to discover if sailing straight north might be a more direct route, which was also ice-free.

In her book about the three expeditions, *Icebound*, Andrea Pitzer writes that the Dutch learned from the experience of the first two expeditions: The first voyage was plagued by discontent and enmity among the crew, and seven men died on the second expedition—two were killed by polar bears and five were hanged for mutiny.[21] Given what might happen on the third expedition, only unmarried men were allowed to sign up for the crew.

The boat that Barentsz was navigating got frozen into the ice. It was only when I sailed to the Arctic Ocean myself that I realized that you stood no chance if you were in a boat without a motor and that did not have a hull suitable for the conditions. A ship can freeze into the ice slowly, over a few days, but it can also happen much faster. If there are extreme conditions with icebergs, chunks of ice, and ice slush drifting around the boat in a storm, it can happen within minutes. Because of their size, icebergs and lumps of ice are most dangerous when they are drifting toward you. But it is the almost invisible ice slush, which is perhaps what Pytheas described from his voyage, that is most treacherous. The slush is made up of ice clumps of varying sizes that float on the surface of the water. It is extremely difficult to sail or row through the slush. From a distance, it may look easy, but once the boat is in the middle of the ice it becomes hard to move forward, or to steer to the right or left, and you may be forced to spend an extra year in the Arctic.[22]

Barentsz and the fifteen other crew members had to winter in a wooden cabin that they built which measured no more than 25.6 × 18 feet. They gave the cabin the optimistic name Het Behouden Huys, which roughly translates as "The Safe House." Most of the men got scurvy. The symptoms include fatigue and breathing difficulties, aching muscles and joints, bleeding gums, and the loss of teeth. It can also result in healed wounds reopening and lead to brain or heart aneurysms and death. If you are healthy and strong to begin with, you can normally manage without vitamin C for about six months, but the illness will eventually catch up with you.

It is likely that the men survived because they were good hunters and so were able to eat Arctic fox meat, which contains traces of vitamin C. It is interesting to note just how little they knew: There was no awareness of the link between vitamin C and scurvy. They had no idea how to determine which longitude they were on and so were never certain of how far east or west they were. The connection between all the bacteria on board the cramped boat, where hardly anyone washed themselves, and diseases was unknown. And the Dutchmen expected to find the Arctic Ocean to be partially ice-free to the North Pole and on to the other side of the world.

The nearby ship was periodically overrun by polar bears and a considerable portion of the account from that winter is about polar bear attacks. The snow lay so thick and high around the cabin that the men sometimes had to shimmy up and down the chimney to get in and out, and polar bears tried to break the roof in an attempt to get in. The crew ate polar bear meat and experienced something that many an expedition member has experienced in relation to food: It might not taste very good to begin with, but as time passes, it tastes better and better the hungrier you get.

One evening, the crew shared the liver of a polar bear they had just shot. If you eat the meat as soon as the bear has been killed, it is quite tasty, but as the animals are so big, it is generally only possible to eat a small portion of the meat before it freezes. So the meat that most polar explorers have eaten is untenderized and tough as a result of having been frozen and defrosted several times. The Dutchmen relished the fresh liver, but did not know that polar bear liver contains enough vitamin A to kill humans. Everyone got sick—they threw up, had intense headaches, were drowsy, their skin peeled off, and three of them almost died.

On January 24, 1597, two of the men saw the sun appear in the sky, two weeks earlier than it should have according to Barentsz's calculations. Gerrit de Veer, one of the crew, made a note of the observation. It is unlikely that the Dutch were the first to witness this seeming permanent sunrise—Russians, and possibly Vikings, had hunted on and around the island for some 500 years—but de Veer is probably the first to document it. When they returned to the darkness of the cabin, no one believed them, and they were accused of lying. Everyone thought it was impossible to see the sun before it was above the horizon. Three days later the entire crew witnessed it. But they were still not believed on their return to Amsterdam.

Many years later, it was proved that the sun in the Arctic can be seen in the sky even when it is still below the horizon. What they saw was an astronomical deviation, a mirage of the sun that had not yet risen. The phenomenon is called the Novaya Zemlya effect, after their experience on the island.

THE SHIP REMAINED frozen in the ice. In June, desperate, they gave up waiting for the ice to release it, and decided to sail and row south over the water that has since been called the Barents Sea, in two small boats. I have crossed the same sea in a sailing boat and the water always looks black, both when calm and when stormy. Other seas and oceans are countless shades of blue, sometimes turquoise and green, but the Barents Sea is so rich in biomass that the light from the sun is absorbed in the cold water and simply disappears. The only time I have seen small parts of the black, monochrome sea change color is when there is a strong wind blowing, and the tops of the waves are white.

The galley boy died before they left their small cabin, and three more men perished on the journey. Barentsz was one of them. He had been so ill before they left, he was pulled on a sled down to the boat and died on board. It is unclear whether he was buried on land or at sea. Seven weeks later, those of the crew who survived arrived at the Kola Peninsula.

Meanwhile, by the summer of 1597, everyone in the Netherlands had given up hope that anyone had survived that expedition. More than a year had passed, so no rescue operation had been mounted. The survivors happened to meet a Dutch boat sailing by the Kola Peninsula and were saved. It was pure luck.

On his return from the South Pole, and long before he traveled to the North Pole, Roald Amundsen wrote that we are the creators of our own luck: "Victory awaits the man who has everything—some people call it luck. Defeat is guaranteed for the man who has neglected to take necessary precautions in time; this is called bad luck." In my experience, this is true. But you cannot prepare yourself for every eventuality, so as a polar explorer, you sometimes need pure luck.

Barentsz may have lacked the extra luck to survive, but he had something else. His failure was apparently more spectacular and heroic than that of other men and previous expeditions. There is something spellbinding about a skilled navigator and cartographer: everyone's fate depends on them, they know better than anyone where the ship is and which course it should take, and they draw maps of the areas they sail through for posterity. When Barentsz's crew returned to the Netherlands, the surviving members told tales of Barentsz's daring, of his fights with polar bears, how he saved the lives of fellow crew members, and listened to everyone. The expedition members may not have reached their geographical goal, but they were successful in another way: they became heroes, especially Barentsz.

Pitzer writes in her book that Barentsz met "wonders and terrors without understanding most of the forces at play in his universe" and he failed "in nearly every way." Yet the expedition awakened real interest in the Arctic. They also produced the first maps of parts of Svalbard and Novaya Zemlya.

It was perhaps Barentsz who first inspired the image of the heroic, suffering, and preferably dying polar explorer that lives on among polar explorers to this day. Only five years after the expedition, Shakespeare wrote about the cold they had experienced in *Twelfth Night*: "You are now sailed into the north of my l'dy's opinion; where you will hang like an icicle on a Dutch'an's beard."

In the nineteenth century, the English coined the phrase *heroic failure* and I have tried to find the same expression in other languages, without success. The expression encapsulates the idea that it is permissible to fail when striving to achieve a great goal, if you fail in a noble and painful way, or have sacrificed yourself as a martyr to the honor of your country. If the story is told with due respect and glory, it is often more thrilling and moving for an audience if the explorer dies or nearly dies in the process, rather than succeeds. Such was Barentsz's fate. Perhaps there is nothing more gratifying than a dead hero who was left behind in the Arctic.

THE MUSCOVY COMPANY'S agreement with Russia continued to be profitable. The company's original ambition of finding a new route to the Pacific Ocean, thus circumventing the Tordesillas Treaty, lived on into the seventeenth century. The company financed several expeditions north, but none came close to discovering an eastern or western passage. The only thing they proved was how difficult such an expedition was.

In 1607, the company changed tack and contracted the Englishman Henry Hudson (*b.* 1565; disappeared 1611) to sail an even more northerly route, over the North Pole or around Svalbard and then east. Hudson reached Greenland and then Svalbard, and thus set the record for traveling farthest north, or as it says in the foreword to Robert Peary's account of the North Pole, "advanced the eye of man to 80° 23′." He was only 3,608 miles from the Pole. (Hudson is remembered today as he gave his own name to both Hudson Bay in Canada and the Hudson River in New York.) When he sailed up the Hudson River, with what is now Manhattan on the starboard side, in search of the Northwest Passage, he was in fact not serving England, but the Netherlands. I find it strange that one person could work for several countries—countries that competed for the same trade routes, no less. But it was not unusual for renowned navigators and captains to sell their services to countries other than their homeland, as Columbus and Magellan did. Thanks to Hudson's discovery, the Netherlands had a territorial claim on Manhattan and in 1625 established the town of New Amsterdam there, as the capital of

New Netherland. It was not until 1664 that the town was renamed New York.

In 1610, Hudson set off on his final expedition, under the English flag, once again with the aim of finding the Northwest Passage. Following a winter in Hudson Bay, his men mutinied. Hudson was abandoned at sea in a small, open boat, together with his son and seven of the crew, who were either loyal to Hudson or too sick to flee. Those who were strong enough to row grabbed the oars and tried to catch up with the ship, but didn't make it, and the nine men were never seen again.

In the decades after Henry Hudson's expeditions, explorers continued to travel west, east, and south, but rarely north. It was not until the 1670s that the dream of the North Pole was given life again.

3

THE POWER OF THE UNKNOWN

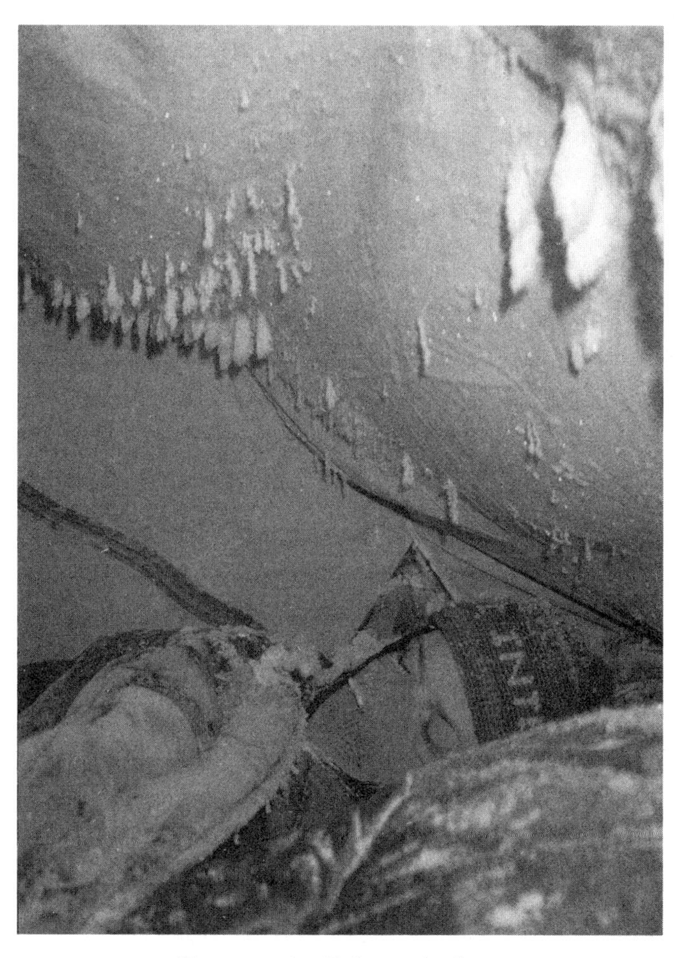

Frost covering Erling as he sleeps.

A Feminist Icon at the North Pole

The myth of the North Pole persisted, without anyone trying to travel there again. American explorer Robert Peary played on this in 1909, when he said that he had been to the North Pole and beyond. The North Pole was a place of magic. A world beyond our world, where no living being had been. But if you succeeded in getting there, or even close, you would be transformed, and move from our world to a parallel world with different natural laws.

It was not unusual to think that a journey to the North Pole was about more than simply getting there. You could cross an unknown border, travel "beyond the Pole" or "into the Pole." The pioneering astronomer Edmund Halley (1656–1742), whom the comet was named after, claimed in 1692 that the earth was hollow. If you traveled far enough north, you would continue down into the earth itself.[1] According to Halley, there were people living at the center of the earth, and the key to the origins of the Northern Lights was also to be found there. Their source was gasses that escaped from the earth's center at the North Pole. In Thomas Pynchon's novel *Mason & Dixon*, which is set in the late eighteenth century, it is not possible to reach the North Pole, as the polar point is suspended in the air. Pynchon writes that the surface of the earth starts to slope downward at 80° N, and anyone who ventures beyond this point will end up in the center of the earth, just as Halley anticipated.

While Halley was trying to solve several of the mysteries of the North Pole, the first woman in the history of the North Pole appeared. The philosopher and scientist Margaret Cavendish, Duchess of Newcastle upon Tyne (1623–73), is the first person to write a novel set at the North Pole. *The Blazing World* builds on existing myths of the

North Pole, and uses the North Pole as a tool in the fight for a woman's right to fulfill her potential.[2] Cavendish was a passionate feminist, and lived at a time when a woman could not be either a scientist or a philosopher. She used whatever opportunity arose in life to show how ridiculous gender roles were. She bowed rather than curtsied, and wore men's clothes that she sewed herself and a velvet hat.

The novel was a kind of prototype of feminist science fiction. A seafaring man falls *extremely* (the author's word) in love with a beautiful woman. He does not waste any time courting her but instead kidnaps her. With her on board his ship, he feels he is the world's happiest man. In more traditional stories, I guess they would "live happily ever after," but Cavendish constructs a far more entertaining narrative. The ship gets caught in a terrible storm and is "carried as swift as an Arrow out of a Bow, toward the North-pole." No one on board is prepared for the cold and everyone, except the woman, freezes to death.

The woman is saved by some bearlike creatures who take her to the Blazing World of the title, "the very end or point of the Pole of that World, but even to another Pole of another World." The capital of the empire is Paradise, and parts of the city are built in gold. When she explores Paradise, she is apparently relieved to discover that the architecture is reminiscent of Roman times and pleased that the city is devoid of contemporary architecture. Not only is Paradise a city of classical architecture, but it is also where the Garden of Eden was once found. She is confronted by the locals: What happened to Adam? No one has seen him since he ran away, and they assume he must have gone to the world that she comes from.

The emperor of the Blazing World believes she is a goddess. They fall in love and get married, and she is given "absolute power to rule and govern all that World."

For seventeenth-century readers, the idea that a woman could rule was radical, even though Elizabeth I had ruled with absolute power only a generation before. It was equally radical for a woman to publish a novel under her own name. Women were denied the right to education, and while it was acceptable to write short, preferably religious texts, it was not deemed proper for women to write philosophical papers or novels.

In Cavendish's imaginings, at the Pole, beyond the North Pole, there are no weapons, and eternal peace prevails. The inhabitants walk upright and behave like people, but have other characteristics that are more akin to bears, lice, fish, birds, or snakes. To her delight, she discovers that these creatures are wiser than the men in her own world.

Liberated from chauvinistic English society, the protagonist can be herself and decides to write what she calls her own kabbalah. She has great ambitions and her initial wish is to make spiritual contact with Aristotle and Plato, so they can help her write the book, but she is advised against this. Old, wise men are "so wedded to their own opinions, that they would never have the patience." She then wants to get in touch with the souls of Galileo Galilei, Thomas More, and Thomas Hobbes, so they can help her with her book. But she is advised against this too, since they are so "self-conceited, that they would scorn to be Scribes to a Woman."

Toward the end of the novel, the empress considers leaving the kingdom at the North Pole and going back to her old world. She imagines a return not dissimilar to the second coming of Jesus Christ on Judgment Day: a bright, shining flame suddenly lights up the world, and everyone is afraid, then all falls dark, a choir starts to sing, and only the empress can be seen "in her Garments of Light." This description angered the Church, as it believed a woman could not perform miracles, and even north of the North Pole women should be subservient to men.

The Commander, the Baron, and the Lord

The tragic fate of polar explorers Willem Barentsz and Henry Hudson led to a waning interest in expeditions to the North Pole. But as the two explorers and other trappers had reported a wealth of wildlife along the edge of the pack ice, there was still considerable activity in the Arctic, mainly in connection with the hunting of walrus and whales. Some summers, as many as 300 ships and 15,000 men traveled north, and several thousand used Svalbard as a base. They built cabins and boiled blubber. At other times the ice was so compacted farther south that they could not reach land. The shores of the islands west of Storfjorden are still white with walrus and whale skulls and bones. A friend who sailed to one of the islands thought it was a white sand beach, until he got closer.[3]

A number of these whaling ships sailed as far north as polar expeditions manned by scientists and explorers. The whalers made discoveries, drew maps, and logged the temperature, winds, and currents. Sometimes conditions were such that they were forced to winter there. Trappers and hunters starved and froze to death, were killed by polar bears, and drowned, but there was little interest in their exploits, and I have never heard of a whaler being given a medal, or becoming as famous as a king or president, as was the case for polar explorers. To this day, there are countless unvisited graves on Svalbard holding forgotten whalers.

FOR A SHORT period in the 1600s, interest in the North Pole flared up again in England, thanks to the publication of a document called "Transactions of the Royal Society of London 1675," which stated that a ship had sailed "several hundred leagues" northeast of Novaya Zemlya and that the ocean there was ice-free. A league is equivalent to 3.4 miles, so the ship would have been in the proximity of the North Pole.

This was, of course, entirely untrue. But it has been difficult to verify such statements until recently. Captain John Wood managed to persuade King Charles II, the Duke of York and the Admiralty to take the opportunity to sponsor a voyage to the North Pole. As soon as it reached the North Pole, the expedition would then carry on south to the other side of the world. The fate that had befallen Hudson and Barentsz had seemingly been forgotten.

Support for Wood's expedition was inspired by a hunger for empire building, as well as the usual desire to find the Northeast Passage and new trade routes. The English had already occupied Ireland and colonized a number of islands in the Caribbean and parts of North America. Wood was given responsibility for the expedition, which was formed of two ships: *Speedwell* and *Prosperous*. They left England in May 1676, and on June 26 the same year they sighted an island that they believed was Novaya Zemlya. The exact position that Wood gave is unclear, but three days later *Speedwell* ran aground and sank. Everyone managed to reach shore, but *Prosperous* had gotten lost in the fog and did not find the survivors until July 8.

NEARLY ONE HUNDRED years later, in 1773, the Royal Society concluded once again that it should be possible to sail to the North Pole. The reason for this renewed hope was that yet another ship had reported partially ice-free water in the vicinity of the North Pole.

But this time there was a different motivation for an expedition. The power play between Spain, Portugal, and Britain had changed, and the British had secured alternative trade routes to the Pacific

Ocean. The driving force now was a northerly expansion of the British Empire. It was the year of the Boston Tea Party, a protest against the taxes introduced by Great Britain on the American colonies, and the empire was in danger of losing them. Great Britain still controlled Ireland, some islands in the Caribbean, and an increasing share of India. (Australia was not colonized until 1788.) The aim was to demonstrate Britain's superiority in the unknown and unexplored world. Personally I am fascinated by the sense of adventure and courage that they showed.

The British also wanted to carry out research in the Arctic. Isaac Newton's seminal work *Principia* was published in 1687 and had made northbound expeditions pertinent for scientific reasons. Newton claimed that gravity determined the orbit of the planets and that there was an "excess of gravity in these northern places." If it was true that gravity was stronger at the polar points, then the world could not be completely round after all, he said, but rather, it flattened out around the polar points. We now know that Newton was right, even though he was unable to prove it. He had access to extensive material from other people's travels closer to the Equator, but could make "no observations myself." There was a need for comparable observations from the north to determine the shape of the earth, as well as the size of the solar system.

Constantine Phipps, latterly Baron Mulgrave (1744–92), was chosen by the Admiralty to lead the 1773 expedition. Phipps had left the family home to join the navy at the age of fifteen. He had already fought in several wars, including the American War of Independence.

The appetite for exploration in Great Britain was voracious. It was hard not to be tempted to give it a go, Phipps wrote, because he lived at a time when "the public mind was elated by the brilliant discoveries" of James Cook. Commander James Cook was the most famous explorer and scientist in the Western world. He succeeded in two of Britain's great international ambitions: to expand the empire and to be a leader in science. On his first expedition, he had sailed around the world via Cape Horn to observe the transit of Venus—when the planet passes between the sun and the earth—in Tahiti on June 3, 1769. The phenomenon always happens in pairs, and the first occurrence had been in 1761, and the two before that in 1631 and 1639. The next would be 1769, and then not again for another hundred years or more. So that day in June 1769 was the last opportunity to observe this transit in their lifetime.

I witnessed the transit of Venus myself on June 8, 2004, standing in Hyde Park in London. No living human had seen it before millions did that day. When I looked up, I saw a tiny black dot moving horizontally across the bottom half of the sun.

PHIPPS WAS GIVEN two war ships by the British Admiralty, HMS *Racehorse* and HMS *Carcass*, both specially adapted for the ice. Members of the expedition had been selected to study zoology, magnetism, time, whales, cartography, and astronomy, and the expedition has often been cited as the first scientific Arctic expedition.

The Admiralty's instructions to Phipps were clear and optimistic: Over the course of the summer they were to sail between "Spitzbergen and Greenland, proceed up to the North Pole, or as far toward it as you are able, carefully avoiding the errors of former navigators by keeping in the open sea, and as nearly upon a meridian to the said Pole as the ice or other obstructions you meet with will admit of." The Admiralty were also clear that Phipps should not carry on toward the Bering Strait but rather sail straight home.

King George III supported the expedition, and Phipps certainly did his best to pay his dues on his return. The first five pages of his book, *A Voyage toward the North Pole*, are a long and unashamed tribute to the king. Once Phipps is done with praising him, he gives an account of the research carried out.

The expedition measured the distance between the moon and the sun, and the moon and Jupiter, and with the help of a pendulum measured the force of gravity. They sailed as far as Sjuøyane ("Seven Islands"), the most northerly group of islands in the archipelago of Svalbard, thereby setting a record for having traveled farthest north. The previous record was set by Henry Hudson 165 years previously, if not Willem Barentsz, though there were Dutch and English whalers in the sixteenth and seventeenth centuries who claimed to have reached even farther north.

Phipps is often cited as the first modern explorer to describe a polar bear, which may well be true. Phipps's account was retold in books, speeches, drawings, and paintings. These stories helped inspire the wave of Romanticism that broke over England toward the end of the 1700s, of which Mary Shelley and her novel, *Frankenstein*, are a part.

One of the main reasons why the Phipps expedition is still remembered is that a fifteen-year-old called Horatio Nelson was part of the crew. Legend has it that he was stalked by a polar bear that wanted to eat him. His musket misfired, so he grabbed the gun by the barrel and swung the butt at the polar bear.

While many are cynical as to how true the story is, I think it is feasible that he used the gun to batter the bear. If the bear was already full, it may have given up and slunk off, especially if the butt hit it on the snout, but if the bear had been hungry, Nelson would not have stood a chance. The story was embellished year after year, and in a famous painting from 1806, entitled *Nelson and the Bear*, which was reproduced and distributed widely, he is portrayed in combat with the bear. They are each standing on an ice floe with only a narrow gap between them, staring each other in the eye. The sky is dark gray, but the area behind Nelson is illuminated, as in religious paintings. According to legend, Nelson told the captain that he wanted to kill the bear, because "I wished, Sir, to get the skin for my father."

The Russians Look North

"Russia is the country with the longest coastline to the Arctic Ocean. This is the area where Russia's honor and navigation can grow, and newly discovered lands can become Russian." So wrote the Russian astronomer and poet Mikhail Lomonosov (1711–65). He compared exploration and research in the north with the wars fought by Russia elsewhere. To win a small area of land, Russia was willing to send entire armies to their death, whereas to the north, with all its potential, Russia could spare "only around a hundred men."[4]

Then, in the eighteenth century, Tsar Peter the Great set his sights north. He wanted to build an empire and to explore the possibility of new sea routes between east and west. But the head of the largest country in the northern hemisphere had no ambitions to reach the North Pole. Perhaps it would be hard to convince a Russian that the Arctic Ocean was ice-free. They knew better, unlike the British. Russian whalers had in previous centuries spoken of an ice edge that moved southward and northward depending on the weather and the seasons, and the pack ice farthest north.

The focus of Russia's exploration and colonization of the Arctic was to find a possible Northeast Passage, from Norway in the far west, east along the coast to what later became known as the Bering Strait, and from there south along the Pacific coast.

The St. Petersburg Academy of Sciences became an important

driving force for Arctic research and the expansion of the empire. Tsar Peter the Great had taken the initiative to establish the academy. His ambition was to challenge the western European perception of Russians as "barbarians who disregard science" by becoming pioneers.[5]

The Danish cartographer and explorer Vitus Bering was commissioned by the tsar to lead the First Kamchatka Expedition. Bering and his team of 600 left St. Petersburg in 1725. They wanted to map the areas to the north and east of the Kamchatka Peninsula, to "search for a place where that land might be joined to America." It took the expedition three years just to cross Russia from west to east.

The tsar's primary concern was to control the fur trade, develop relations with Japan and America, and form the basis for a territorial claim in the Arctic.

When Bering reached Kamchatka, he procured a boat and sailed north. He reached the strait and could then send confirmation to his sponsor that there was indeed a passage between the Arctic Ocean and the Pacific Ocean. He looked out for Alaska, but never saw it.

By the time Bering returned to St. Petersburg in 1733, Tsar Peter had died, but his successor, Anna of Russia, was extremely pleased with the expedition results. With the support of the Academy of Sciences, she initiated the Second Kamchatka Expedition to map the land, sea, and islands in detail. In 1733, Bering set off with a company of 3,000 people. They reached the strait and crossed it to become, perhaps, the first Europeans in Alaska. Bering died at forty-one, in all probability from scurvy, on an island that was as yet unnamed. Later it was given the name Bering Island in his honor.

The success of Bering's expeditions laid the foundation for several more. The increasing presence of British, French, and Spanish vessels around the Bering Strait worried Catherine the Great. In 1779, after several failed attempts, a Spanish expedition reached Alaska, which, according to Spain, was a Spanish possession, based on the Tordesillas Treaty, which had divided the world vertically from pole to pole.

JAMES COOK ARRIVED at the Bering Strait in 1778, on his third expedition, having set sail from Plymouth in England in 1776. He'd sailed south around Africa, then east across the Indian Ocean to New Zealand, then north across the Pacific Ocean to the west coast of America, and finally into the Bering Strait. He left messages in bottles on land that said, in Latin, that King George III's ships had been there.[6]

Cook had retired after his second great expedition. He had more than enough money, and lived in a house in Greenwich, outside London, with his five children and pregnant wife. But he was bored and was the first to admit that he found his new life limiting. As he wrote to a friend, the southern hemisphere had barely been big enough to keep him occupied. So when the Admiralty asked him to lead yet another expedition, he did not hesitate in saying yes.

Two goals of the expedition were scientific research and potential colonization. But on board his ship, HMS *Resolution*, Cook also had a third, secret order from the Admiralty. It was indeed secret from even the rest of the crew and other officers until a year into the voyage. The expedition was not to return to Great Britain via the usual route, but instead, in Cook's words, "to make my passage home by the North Pole." The voyage north, from the Pacific Ocean, over the top of the earth, to the Atlantic Ocean and back to Britain on the other side of the North Pole, was to be Cook's last great triumph.

The British were still convinced that the Arctic Ocean was ice-free, even though Constantine Phipps had been stopped by pack ice more than 621 miles from the North Pole only a few years earlier. The Admiralty received relatively regular reports of ice-free conditions in the far north. The two British ships were therefore not adapted to deal with the pack ice. An English report, *Tracts on the Probability of Reaching the North Pole*, published in 1775, mentions a Dutch captain who had almost reached the North Pole.[7] According to the captain, there was no ice and it was so warm there that he did not need a jacket. The author of the report added that it was as warm as summer in Amsterdam at the North Pole.[8] I have no idea what spawns such stories, but I suspect that the Dutchman who was cited as a witness was pulling the wool over the Englishmen's eyes. And if that is the case, he succeeded.

In summer 1778, Cook managed to get beyond the Bering Strait, but the expedition was stopped by ice just off Alaska. To the despair of the crew, he chose not to turn and sail south, but rather to follow the ice edge west to Siberia, in the hope of finding an opening passage going north. Cook was surprised to discover that the ice was compact all the way. The mist also took him off guard. It was so thick for a period that Cook had to sail blind, and the masts and rigging of HMS *Resolution* were soon cased in ice, as a result of the damp air. The ice added weight to the ship and her center of gravity was soon so low that the crew feared she might keel over.

After Cook was murdered[9] on Hawaii on February 14, 1779, when the British tried to kidnap King Kalaniʻōpuʻu-a-Kaiamamao, the new leader of the expedition attempted another "attack upon the North Pole," but the ships were stopped by ice even farther south than where they had turned the year before.[10]

Russia, France, and Spain continued to explore the northwest area of the Pacific Ocean. Catherine the Great sent another expedition to explore and map the coast of Siberia and Alaska, and to claim sovereignty over any new islands, or to prevent anyone else from doing so. And it would seem that the tsarina chose the right strategy, as these territories, with the exception of Alaska, still belong to Russia. Spain, France, and Britian failed to gain any new territory.[11]

The Sublime and Honorable Suffering

The story of Nelson's encounter with a polar bear on Constantine Phipps's expedition portrayed in the painting *Nelson and the Bear*—as a close, life-and-death encounter—was an omen of what was to come in polar history. Throughout the nineteenth century, it was no longer enough for a polar explorer simply to recount what had happened on the expedition. Polar literature from this period is rarely about the challenges of drying underclothes at minus forty or working out where north is when the compass needle is pointing west. Daily tasks were not enough. Nor were beautiful descriptions of nature. Readers and audiences expected more.

In his book *A Philosophical Inquiry into the Origin of Our Ideas of the Sublime and Beautiful* (1757), the English philosopher Edmund Burke wrote that the aim of the sublime is that "it is productive of the strongest emotion which the mind is capable of feeling." Burke was a Romantic, and his thoughts on the sublime formed part of the ideological and aesthetic basis for how dramatic challenges in the Arctic should be faced and then shared. The men on these expeditions allowed themselves to be swept along by the ideal. The more you suffered from the cold, hunger, and weather, the more honorable it was.

Burke wrote the book when he was twenty-four, after he had tried to study law and given up. He writes that the sublime surpasses the beautiful, and the source of the audience's experience of the sublime is anything that is suited to "excite the ideas of pain, and danger, that is to say, whatever is in any sort terrible . . . or operates in a manner analogous to terror."[12] It is also important to differentiate between those who listen and those who read, and the storytellers themselves. As Burke writes, "When danger or pain press too nearly, they are incapable of giving any delight, and are simply terrible; but at certain distances, and with certain modifications, they may be, and they are, delightful."

The experiences of pain and terror—the wind, cold, and ice that are out of the explorer's control—in turn elicited feelings of fear and hopelessness in the reader. A bleak, dark, and gothic Arctic landscape illuminated by the moon or the Northern Lights was an ideal place for the sublime. There, the polar explorer you are reading or hearing about is faced with multiple trials in the form of frostbite, longing, and disease. And not only that, the explorer, who showed great courage and did not get angry, but rather accepted reality and handled his suffering with stoicism, demonstrates a profound moral superiority.

IF AN EXPERIENCE on the ice is to be elevated to the sublime, it must enact some kind of transformation in the minds of the readers or audience. True sublimity must, according to Immanuel Kant, "be sought only in the mind of the one who judges, not in the object in nature, the judging of which occasions this disposition in it."[13] To see a sunset or observe the countless, beautiful shades of green in a forest is wonderful, and it is precisely these experiences that give us a sense of connection to our world.

But the sublime was an entirely different story. According to Kant, we humans do not belong in the sublime world, because the sublime experience can often be so overwhelming that it is hard to process. Our powers of imagination break down when it is too dramatic, we feel that we are different from the people and nature we are reading

A theater poster from a Broadway melodrama in 1896 emphasizes the "Romantic Spectacle" of the North Pole.

about. As rational creatures, we feel weak in the face of the sublime, and therefore, according to Edmund Burke, we fear it.

Kant went for a walk every day, but he never left his hometown of Königsberg on the Baltic Sea, now known as Kaliningrad in Russia, from the day he was born to the day he died. And yet it is easy for a polar explorer to understand his thinking. When we started our trek to the North Pole in 1990, it was too cold for me to philosophize. But as the cold eased and it was no longer 40 or 100 degrees below zero, but closer to –5 or –70 degrees, I started to appreciate the landscape. I could think about the frostbite, exhaustion, and headwinds we had faced and the dangerous, thin ice we had crossed with a mixture of fear and exhilaration. And the joy of feeling warm after having been freezing, full after having been starving, and rested after having been exhausted is, I believe, the original source of gratitude.

The German Romantics described nature as God's temple and idealized the gravity of the north on people's psyche, and the brutal nature they would encounter there. Caspar David Friedrich is the one who perhaps best illustrates the power of ice. In his painting *Das Eismeer*, three quarters of the picture is huge ice floes pressing against each other and towering up into the air, and above the ice, the sky. Slightly off-center, there is a ship that is being crushed by the ice. The shipwreck and unanswered question as to what might have happened to the crew crystallize Kant's theory of the sublime. A sublime painting cannot tell the whole story; it has to include something untold that indicates drama, even death.

The first time I saw the painting in real life in the Hamburger Kunsthalle, I was astounded by how Friedrich depicts the Arctic Ocean, seen from the south.[14] The ice floes that are rammed together to create the hummocks that crush the boat are very similar to those I have seen. Even fixed in paint, it is as though the ice is moving, and you can hear the ship's timbers being crushed. I scoured the painting for people, living or dead, but found none, and was left with the impression that everyone had perished in a sublime way along with the ship.

FOR BURKE, THE law school dropout, an ideal, sublime day might have been one that was stormy, perishingly cold, weighed down with snow, when a man had to persevere or freeze to death, or survive a polar bear attack.

But the truth is, when you find yourself in those conditions, it is impossible to contemplate the ideas of Kant and Burke. On days when the temperature sinks to −60°F, you only have one thought in your head: It is cold. You have to get warm, to stay alive. But other people's pain and danger make for a thrilling yarn. The sublime "surpass[es] every possible phenomenal, empirical experience," according to the philosopher Slavoj Žižek. The person listening, reading, or seeing experiences a vicarious taste of the life-or-death drama, without having to suffer, and that is one of the reasons why polar expeditions remain so popular.[15]

Frankenstein's Journey to the North Pole

Not many people would think of the scientist Victor Frankenstein in Mary Shelley's novel *Frankenstein* as a polar explorer. The name is more often associated with hubris and the possibility of creating artificial life, as well as, today, artificial intelligence. And yet his journey to the North Pole is the most important journey of his life.

To my knowledge, Doctor Frankenstein was the first fictional character to express the dilemma that all North Pole explorers face:

Should I carry on north? Or should I give up? All the way, he is convinced he should press on northward, even when he is close to death. Just before death takes him, he realizes he will never reach the North Pole.

The novel was published in 1818. This was nine years before the Englishman William Parry (1790–1855) sailed toward the North Pole on an expedition that would mark the start of the rise of polar heroes.

Frankenstein's two extreme acts, first to create a human being and then to travel to the North Pole, are both based on arrogance. He wants to outdo nature by creating a beautiful humanlike variant, for the good of mankind, with the help of biology and what we now know as artificial intelligence. His failure to create something beautiful is obvious, but his success in creating something far more complex—a living being—perhaps less so. Frankenstein creates a living being that is dependent on loving and being loved, and is torn between the same virtues and sins that humankind has grappled with since time immemorial: trust, betrayal, friendship, revenge, honor. Everyone who meets Frankenstein's nameless being is repulsed by him because he looks so different. He is even betrayed by his own creator, Victor Frankenstein. Frankenstein's creation is neither a person nor a monster at the outset, but disappointment pushes him into being a vengeful monster who kills Frankenstein's brother and best friend, and eventually also his bride on their wedding night.

The monster flees to the North Pole. Frankenstein wants revenge following the murder of his loved ones, and, fueled by arrogance, recruits a team of dogs and pursues the monster that he himself created. Everyone in this novel suffers.

Far out in the Arctic Ocean, the exhausted and frozen Frankenstein meets a ship that is sailing to the North Pole. He is about to freeze to death but will only let himself be rescued on one condition:

that the ship does not turn. The captain, R. Walton, promises not to turn. Walton's ambition is to be the first person to reach the North Pole, to "tread a land never before imprinted by the foot of man" and discover the force that attracts the compass needle. He had always imagined the North Pole as a happy and beautiful place, but as he sailed north he understood that the Arctic is cold and barren, and toughened himself by freezing, going hungry and thirsty, and getting by on very little sleep. Shelley's description of Walton's romantic dreams of the North Pole reminds me of the Norwegian polar explorer, Roald Amundsen. Around fifty years after the book was published, the boy Roald Amundsen slept with his window open through the winter to toughen himself up. He had already decided that he would be the first to sail the Northwest Passage and had started his preparations.[16]

Frankenstein was weak and bedridden. It is September and the polar night is closing in and the boat is locked in the pack ice. The crew realize that to carry on north would in effect be collective suicide. They threaten mutiny if the ship does not turn and head south again as soon as the boat is released from the ice. When Frankenstein hears about this, he musters all his strength to give a speech about the importance of behaving like a man and why it would be honorable to carry on: "because it was full of dangers and terror; because at every new incident your fortitude was to be called forth and your courage exhibited; because danger and death surrounded it . . . You were hereafter to be hailed as the benefactors of your species."

Victor Frankenstein closes his speech by comparing the men's hearts with the ice. The ice is weak compared to their hearts, because "it is mutable, and cannot withstand you" and everyone on board can return home as heroes, and not cowards.

Walton decides to turn back all the same, and writes a letter to his sister telling that all hope of honor and good deeds have been "blasted by cowardice and indecision."

I do not see this as cowardice or indecision but think, rather, that Walton has understood something that Frankenstein never did, a detail that is Frankenstein's undoing: You cannot and never will win against nature. Nature can benefit humanity, but only as long as we respect it. Shortly before Frankenstein dies in his bunk, it dawns on him that he has made the wrong choices, and he gives Walton some advice, this time on the importance of surrendering. "Seek happiness in tranquillity and avoid ambition, even if it be only the apparently innocent one of distinguishing yourself in science and discoveries."

This is something that many polar explorers have acknowledged when they—like Frankenstein—were exhausted and frozen. Then, when you feel revived again, all memories of the cold, hunger, and danger are suppressed, and the ideals of Frankenstein's speech take over once again. In my experience, the toil, cold, and hunger are remembered with a smile when recalled from a distance. As the saying goes, "All's well that ends well." Seneca wrote that things that are very hard to bear are sweet to remember. No one can take your experiences away from you, unlike anything else you might have.

Frankenstein's final words in the novel are: "I have myself been blasted in these hopes, yet another may succeed." It seems he has learned nothing.

The creature continues his journey to "the most northern extremity of the globe," to gather wood and "ascend my funeral pile triumphantly." What a romantic thought: to light a funeral pile at the top of the world, where he believes the earth is closest to heaven. Edmund Burke would have approved. But the idea that there would be enough wood for a fire demonstrates total ignorance of what the North Pole is actually like.

I am in awe that Shelley was able to describe precisely what it is like to walk over drifting ice in poor conditions, without ever having been anywhere near the Arctic. "The wind arose; the sea roared; and, as with the mighty shock of an earthquake, it split, and cracked with a tremendous and overwhelming sound." Like the theories of Herodotus, Bal Gangadhar Tilak, Henry Corbin, and William F. Warren, and the maps of Ptolemy and Mercator, the book is a paean to people's dreams.

Despite a mixed reception, the novel became a bestseller and was swiftly translated into at least thirty languages. *Frankenstein* piqued people's curiosity about the North Pole and the Arctic: Shelley's story was an indication of how Romanticism, the sublime, and expeditions to the North Pole would go on to mutually inspire one another for decades.

Walton's dream of being first has become a common ambition for polar explorers, as more and more have joined the long ranks of those who failed.

The Polar Bug

Historians tend to divide the history of the world into two periods. The first period is when *Homo sapiens* start to disband, leave East Africa, and head north, west, and east. The most likely cause for this migration was a change in the climate, a dramatic cooling down. This meant we had to work harder and move over increasingly larger areas to find food. When the world then started to warm up again, and the last ice age thawed, *Homo sapiens* followed the receding ice north into Europe and Asia.

In the second period—through the Stone Age, the Bronze Age, the

Iron Age, and the Middle Ages—the movement of people has been more about congregating together. It is an era when explorers made the unknown world much smaller—migration, cultural exchange, trade, agriculture, exploitation, and racism followed in their wake.

Polar explorers diverge from this development and have in many ways continued and bolstered the first tradition, and traveled as far away as possible from other people. There is a reason why we are not called polar tourists.

The dramatic arc of travel accounts often involves people traveling far and wide, only to find the very answers they were looking for when they return home. But not every journey is a journey home. The North Pole explorers seldom had great, or even small, realizations when they finally returned. In contrast to classical heroes such as Gilgamesh, they did not have an epiphany once they achieved their goal. Polar explorers have been bitten by the polar bug, and sometimes polar madness, and need to set off again. T. S. Eliot's poem "Little Gidding" is often read when people want to understand explorers, or to honor or remember them. It is easy to nod in recognition when he says that every end is a new beginning, and we must suffer to move on. He also says that the end goal of exploring is to come to where we began and really "know" our starting place, finally.

THE STORY OF those who felt compelled to reach the North Pole seldom has a happy ending, and more often than not ends in tragedy. That is the nature of the North Pole. Of all the geographical goals that people have struggled to achieve—a sea route to India, China, and the Pacific Ocean, the source of the Nile, the Northwest Passage, the Northeast Passage, the South Pole, Mount Everest, crossing the Pacific—the North Pole is, in my experience, the hardest.

Great Britain's Golden Age

At the start of the nineteenth century, Great Britain was the richest and most powerful nation in the world. The country was a leading naval power and would become a world leader in polar exploration. Three historical circumstances, in particular, laid the foundations for the country's superiority in polar exploration in the decades that followed.

Firstly, for once, there was peace. The Napoleonic Wars had been won; they started in 1802, but it was not until 1815 that Napoleon was finally defeated and exiled to the British-ruled St. Helena in the South Atlantic. Officers, crews, and warships idled listlessly. Pierre Berton writes in his book *Arctic Grail* that the number of men in the navy fell from 140,000 to 19,000.[17]

Britannia needed a new enemy. As Pierre Berton writes, it was neither Spain nor France, nor any other nation, but "the elements themselves" that they turned to, in the form of temperatures as low as 20–60 degrees below freezing, ice, polar bears, wind, and water. But there were other enemies that the British did not anticipate when the idea of the North Pole and finding the Northwest Passage emerged. The days and nights felt colder and more monotonous than they had imagined, the ice was constantly drifting south. There were hunger, lead poisoning, zinc deficiency, scurvy, cannibalism, and insufficient clothing and food suited to the climate and physical exertion.

The willingness to organize expeditions was born from a desire to increase the British Empire. This ambition was the second historical circumstance.

The third historical circumstance was a sudden change in the climate in northern Europe, something that has occurred at irregular intervals, without warning, throughout history. Haraldur Örn Ólafsson, an Icelandic polar explorer, Børge, and I experienced something of the kind in April 2010, when we were trekking across Vatnajökull on Iceland, the second largest glacier in Europe. Below the ice cap of Vatnajökull, there are several volcanoes. As we walked, a huge eruption started below the neighboring glacier, Eyjafjallajökull. The airspace over Europe was filled with ash, which brought nearly all air traffic to a halt. We were in no danger—the force of an eruption goes straight up and does not spread to neighboring volcanoes—but we were reminded of how earlier eruptions had left many parts of Europe deserted.

Events were even more dramatic in 1816, which has since become known as the "year without summer." In April the previous year—two months before the Battle of Waterloo—there had been an enormous eruption from Mount Tambora on the island of Sumbawa, in what is now Indonesia.[18] It has been calculated that the eruption was thousands of times more powerful than that on Eyjafjallajökull. The lava is said to have spewed 130 feet up into the atmosphere. And the eruption was heard 1,600 miles away. A total of one million people are reckoned to have died as a result of the eruption.[19]

One lesser-known consequence of the eruption in Indonesia was that Mary Shelley, who was visiting Lake Geneva in Switzerland, was stuck in a cabin. The change in weather meant it was more tempting to stay in and write *Frankenstein* than to go hiking.

Naturally, many people in Europe were terrified. They thought the world was about to end when summer failed to arrive, and no one knew why. There was a frost in June, and the snow that fell was orange and brown. The corn that had been sown throughout Europe failed to grow. Livestock starved to death. Lord Byron, another writer stuck in the cabin on Lake Geneva, wrote the poem "Darkness" about a dream that turned out not to be a dream: "Morn came and went—and came, and brought no day" when, as he described anecdotally, "the fowls all went to roost at noon and candles had to be lit as at midnight."

THE ENGLISH WHALER William Scoresby (1789–1857) traveled to London in 1816 with the news that the area between Greenland and Svalbard, which was normally covered in ice, was "quite void of ice."[20]

The information that parts of the Arctic Ocean could suddenly be crossed in a sailing boat convinced the Royal Society and Admiralty that the Northwest Passage and North Pole were now within reach. Britain was already a nation with a considerable legacy of exploration, thanks to men like James Cook and Martin Frobisher.

Scoresby was an almost perfect leader for the expedition north. The year before, he had claimed that it should be possible to reach the North Pole in the course of a long summer—with "a probability of success"—with the help of dogs and sleds, or reindeer, if dogs were unavailable.[21] The dogs would have to be acclimatized to Arctic conditions and the sleds should be built in wood, so they were light and flexible, and could be pulled over uneven ice. The mushers would have to be experienced, and he recommended using Inuits. Any expedition would have to start in April or May, while it was still cold, but not too cold. In Scoresby's experience, later in summer the ice was rotten with puddles on the surface and the many open-water channels in the ice made progress difficult by boat or on foot.

Scoresby's recommendations for a potential expedition to the North Pole were excellent.

Scoresby was wrong on only two points, albeit important ones. He was aware of the current that ran south from the North Pole toward Svalbard and Greenland, but he underestimated its strength. As the current south is sometimes faster than you can manage to walk north, it may be impossible to walk across the ice from Svalbard to the North Pole. Certainly, no one has managed to this day. Expeditions that chose to follow this route did not stand a chance. Furthermore, in the experience of most polar explorers, one should set off earlier in the year, preferably the beginning of March, to reach the North Pole before the ice starts to melt in summer.

Scoresby, the whaler, who had racked up 55,000 nautical miles in the Arctic Ocean, said that any notion of ice-free waters around the North Pole was wrong. He claimed that the North Pole was in the middle of an ocean, covered by ice, and that the ocean currents ran freely under the ice between the Bering Strait and Europe. To back this up, he told the story of when he had caught a whale on the European side that had already been shot and injured by a Russian whaler, and the harpoon was still embedded in the creature's flesh, meaning it had swum through clear water.[22] In 1806, when he was working on his father's whaling ship, they had reached 81° 30′ N. This remained the unofficial record for the most northerly position for the next twenty-two years.

Later, whenever he was out whaling, he measured the water temperature and depth, and in 1813 he made an important scientific discovery: The water in the Arctic Ocean is warmer in the depths than on the surface. Herman Melville read Scoreby's trailblazing book *An Account of the Arctic Regions* (1820), about the ocean, the ice, and whaling, and used this knowledge in his 1851 novel, *Moby-Dick.*

Ishmael refers to him specifically in the novel as an authority: "No branch of Zoology is so much involved as that which is entitled Cetology, says Captain Scoresby, A.D. 1820."[23]

Scoresby traveled back to London from his hometown of Whitby in 1817 and offered to lead an expedition to the Arctic. The Royal Society was amenable to the idea, but the Admiralty thought it improper to have a whaler as the expedition leader.

The Admiralty selected its candidates based on contacts, social codes, class, and power, rather than on who was best suited to explore the Arctic. In addition, it was unacceptable to have a leader who was not a naval officer. It is easy to shake our heads at the choice now, but the men in blue uniforms—coattails and all—in the Admiralty offices were children of their time, and were committed to the officers who served them. Yet their preferred candidates all had the same flaw: They had ample seafaring experience from the Napoleonic Wars and other wars fought by Britain, and from expeditions to Africa and North America, but no experience of the Arctic.

THE MOST OBVIOUS reason why British expeditions never reached the North Pole, quite apart from the fact that it is incredibly difficult, is that those responsible in London thought too highly of themselves. They underestimated the importance of preparation and did not listen enough to people like trappers, whalers, and Inuits, who actually know about conditions in the north.

Much of the blame for the lack of preparation has been ascribed to the linguist and geographer John Barrow (1764–1848). Barrow was called the father of Arctic exploration.[24] He held the influential position of Second Secretary to the Admiralty and was responsible for exploration and expeditions. Barrow believed that the Arctic Ocean was ice-free and that the axis running through the planet from the South to the North Pole ended in a mountain at the North Pole. He thought the mountain was most probably formed of dark basalt rock and warm water ran off down the mountain and pushed the ice south. There was a general belief that the North Pole was not unlike the polar point portrayed on Mercator's map.

Barrow told the king, politicians, and public that Britain's honor was at risk over the Arctic. In particular, he highlighted the threat from Russia and the country's investment in new battleships. The Second Secretary maintained that the Russians were looking beyond their own territory, which included Alaska, in an attempt to win the race to be the first to sail through the Northwest Passage. It was a persuasive argument. Another aspect that he played on was that northern expeditions were *peculiarly British.*

One major difference in British and Norwegian polar history is that the Norwegians never had a powerful player like Barrow—someone who decided on behalf of the government who was allowed to go to the North Pole. Not only that, he decided from his office what kind of ship, equipment, food, and medicine the expedition would have, as well as which route it should take, without ever having been in the Arctic himself. If he said no, you would not be part of the expedition. In Norway and the US, no one asked for permission.

In hindsight, it seems obvious, as Fergus Fleming writes in *Barrow's Boys*, that almost everything that could go wrong did go wrong on the expeditions. The orders, the ships, the supplies, the funding, and the methods.

Fleming goes on to say that perhaps never in the history of exploration has a single person "expended so much money and so many lives in pursuit of so desperately pointless a dream." It's absolutely true that Barrow organized the highest number of expeditions that ended in tragedy.

Barrow's greatest shortcoming, and one that he shared with many others, was that he expected British explorers to battle with nature and conquer the North Pole and Northwest Passage in the north and the rivers of Africa in the south. Nature was to be subjugated to humanity's will.

The thought that nature exists for our sake is not new. In the Bible, people are superior to nature and are created to work the land, catch fish, and harness the world, but nature and people belong together. Nature is God's creation, it does not belong to us, so we should be respectful and live in harmony with nature. Aristotle wrote that nature was created so people could harvest from it, and while, according to him, we are unlike anything else in nature, we are still a part of it. It seems that this way of thinking was first questioned at the start of the Enlightenment.

The English philosopher Francis Bacon called for a new worldview. Bacon has been called the first natural philosopher, but whenever I read him, I am more convinced that he spent too much time sitting indoors and too little time walking, looking at the trees, grass, plants, and water. He was skeptical of the theory that people are an organic part of nature. Instead, he helped to develop a new and more tendentious stance, which became popular during the Enlightenment: that people are alienated from nature, because it is so brutal and ruthless, and it must therefore be controlled. Bacon believed that the world was created *for* humans, and we should therefore use science, hard work, and industry to conquer nature. He believed this would not only benefit people in their search for knowledge and power, but also, strangely enough, nature. History has, I believe, proved him wrong on the latter.

Barrow and his contemporaries had the ambition that Britain would conquer the world—conquer in the sense that unknown lands would be mapped and fall under British rule, they would be explored and studied, any fertile lands would be farmed and minerals would be exploited. From our current perspective, it is easy to see this stance as arrogant, but at the time there were very few who criticized this rejection of Aristotle's life philosophy. It was as though Bacon and Barrow forgot that, no matter what the British might accomplish, nature will still be there when we are gone, and that we will never be anything more than a parenthesis in the natural history of the world.

Another weakness shared by Barrow and the Admiralty was that both remained equally convinced of their own excellence and built their theories on the limited science and knowledge of the day, rather than consult more people who really knew the Arctic.

The lack of curiosity about the experience of others ran deep, and is a characteristic that still applies to some British explorers. And I am not saying this out of arrogance: I have good friends who believe in brawn, willpower, and the ability to tolerate pain more than sorting out boots, clothes, food, and transportation before the expedition starts.

THE NORWEGIANS AND Americans had a different attitude and were more curious about the Inuits' ability to survive. Fridtjof Nansen would in all likelihood have died, had he not spent a winter with the Inuits, living as one of them. Robert Peary would never have gotten anywhere near the North Pole if he had not first learned from the Inuits how to drive a dogsled. He later also took several Inuits north with him.

When Børge, Geir, and I were preparing to go to the North Pole, we tried to learn as much as possible from the Inuits. We studied the

Inuits' equipment in detail. Some of their solutions were better suited to Nansen and Peary, who used dogs, than to us, as we were going to pull everything on sleds ourselves. When you are pulling 265 pounds, you will sweat too much in a sealskin parka, even when it is –40°F. Getting wet in the Arctic can be dangerous when it is cold, but also when it is not cold. To our surprise, we noticed that it was not normal for the Inuits to wear hats, no matter how cold it is, only a hood. The hoods are big enough to allow air to circulate around your head and the edge is trimmed with good fur, just slightly farther forward than your face. The fur insulates your face from the wind and driving snow, and the air around your face warms up a few degrees. At the end of the 1980s, there were no manufacturers who made anoraks based on this knowledge, so we had them specially made.

We planned to melt snow on our way north for drinking water and making food. And we would dig up the snow using a spade, as we do in the mountains in Norway in winter. But the Inuits taught us that it is better to use ice. Snow contains more air than ice, and we would use less petrol for every quart of water. That way we saved not only in weight on the pulk, but also the time it would take to melt the ice and make food. The spade we had thought of taking with us was not suitable for hacking ice, so the Inuits advised us to take a small axe. We could then chop the ice into pieces that fitted the pan, which again meant we would save on petrol and time.

It must have been difficult for the first British polar explorers to differentiate between what was saline ice and what was not. The ice in the Arctic Ocean contains as much salt as seawater when it freezes, but over the next one, two, three years the salt is pressed out. This is because salt doesn't freeze, and the ice crystals expel salt into the water, so that when you melt older ice, you get fresh water. Freshwater ice is darker and slightly coarser; saltwater ice is clearer and finer. And if you get it wrong, you get salt water in your glass and food. When you know what to look for, it is easy to tell the difference in daylight. I remember that when we started our trek to the North Pole, it was so dark in the mornings and evenings that we could not see the difference between new and old ice. We had to taste it, and on a couple of occasions my tongue froze to the ice. I then had to warm my tongue and the ice up by breathing in through my nose and out through my mouth, and at the same time use my lips to thaw the ice until it melted and I could pull my tongue free.

Another piece of good advice was to dry mittens by your groin, rather than on your chest, as we were wont to do. Mittens always get wet on polar expeditions, because you sweat in the palm of your hands even when it is cold. Your groin is nearly always warm and the mittens stay there even if you turn over in your sleeping bag. If you gather enough of this kind of experience, the expedition will be safer and simpler.

William Parry and John Franklin

The history of the many attempts to reach the North Pole is also the history of how illusions about a frozen wasteland can lure people to their death while satisfying newspaper readers' growing appetite for drama, suffering, and death.

In Great Britain, there were many who were willing to risk their own and others' lives for the sake of a good story. The two most popular and apparently fearless Romantic heroes were William Parry and John Franklin.

Franklin left home and joined the navy when he was only twelve years old. Before becoming a polar explorer, he sailed around

Australia and fought in the Battle of Trafalgar and the War of 1812 against the US. Franklin appears to be the kind of hero the British wanted. On his first expedition to chart the east coast of North America, he earned himself the nickname "the man who ate his boots." As the moniker implies, when they ran out of food, Franklin got so hungry that he boiled and ate his leather boots—a bit like Charlie Chaplin in the film *The Gold Rush*. Stories like that quickly become legend. But it was not only Franklin who suffered; ten members of the team under his charge died from hunger or froze to death. Later, there were rumors that the boots were not enough, and the crew had resorted to murder and cannibalism. There is little doubt that there were incidents of cannibalism, but whether anyone was actually killed for food remains unclear.[25] Franklin's account of the expedition, which of course makes no mention of cannibalism or poor planning, became a bestseller. The Admiralty was so pleased with the enthusiasm the expedition generated that it sent Franklin off on another Arctic expedition to North Canada.

William Parry joined the navy when he was thirteen, and served for forty-two years. He had originally dreamed of exploring central Africa, but when he was ordered to go to the Arctic, he said the north was just as good.[26] Then, in 1826, he was offered the opportunity to organize an expedition to the North Pole the following summer. According to Parry, the aim was to "complete our knowledge of the surface of the globe."[27]

The North Pole expedition was Parry's fifth trip to the Arctic, and his fourth expedition as leader. On the first expedition he led, in 1819, he got farther west along the Northwest Passage than any other European before him, a record that held until Roald Amundsen completed the entire passage in 1906. He and John Franklin were the two with the most experience leading northern expeditions in the Royal Navy. Parry was so good at writing and talking about his experiences that he also became the most famous and popular explorer in Britain.

Parry's plan to reach the North Pole by sailing to Svalbard was not new (it was similar to Henry Hudson's plan). Pulling sleds from there to the North Pole was not new either (that had been Scoresby's idea). Furthermore, sailing north from Svalbard also meant the expedition had a possible line of retreat, should a boat need repair or be shipwrecked. A few years previously, John Franklin had also sent an application to the Admiralty to lead an expedition to the North Pole using the same route. It was turned down. John Barrow preferred to give authorization to Parry, who was more popular. (Apparently, Barrow discreetly passed Franklin's plan, which he had read and rejected, to Parry and advised him to present it as his own.[28])

It was not Franklin's achievements that made him a hero, rather the fact that he disappeared. He sailed from England on May 19, 1845, with two ships, *Erebus* and *Terror*, with the aim of finding and mapping the Northwest Passage. It was the best planned and equipped expedition to date. Not only did the ships have sails, but they also had steam engines that would keep them moving at four knots if there was no wind. And they had enough food on board for three years. The parts of the ship that were exposed, such as the propeller and helm, were constructed so that they could not be damaged by the ice. The expedition was last seen by Europeans on a ship on July 26, 1845. All 129 men on board died in the course of the next few years, but no one outside the Arctic knew. The uncertainty of what might have happened to John Franklin, his officers, and crew—in that order—generated enormous interest in Europe and the US in the Arctic and the race to reach the North Pole.

THE START OF an expedition to the North Pole in modern times differs enormously from Parry's time. In 1990, like other modern expeditions, Børge and I flew to the northernmost point of the mainland, and the expedition started on the ice from there. The ice, on the other hand, was more or less the same for us as it had been for Parry: vast fields of ice in shades of white, gray, and blue as far as the eye could see, with channels of open water in between. In the winter, the moisture rises up from the channels and freezes into ice crystals in the air. Børge called these ice crystals "silver rain" when they sparkled in the light, before landing on the ice and turning to rime frost. On bigger channels, the ice hung like mist over the open water. And above it all, the sky. Polar explorers often feel that the sky is lower in the far north than anywhere else on earth.

The pack ice varies in thickness, formation, and size from year to year. In 1990, I wrote in my book about our journey, *Nordpolen*, that if climate change continues, future North Pole expeditions will be very different. But I did not anticipate the speed with which this would happen: The steadily warming climate has transformed the Arctic Ocean. The ice that was waiting for us was one to four years old. The hummock formations varied: The newest were high with sharp peaks, whereas the older ones had been rounded and reduced over time as they drifted across the Arctic Ocean. For Børge, the older hummocks looked like rippling waves on an ocean.

The ice no longer has time to reach such a ripe age. It is melting far more quickly. The pack ice that covers the Arctic Ocean now is flatter and, on average, thinner. Obviously, it is easier to cross flatter ice, but because it is thinner, it moves faster, which is exhausting when you are walking against the current.

WILLIAM PARRY'S EXPEDITION started on the Thames, ahead of a long voyage to the edge of the ice. HMS *Hecla*, a full-rigged ship, set sail from Greenwich to Norway and Svalbard on April 4, 1827, then carried on north until they met pack ice and hummocks.

No one on board had appropriate clothing for the climate there, and their equipment was better suited to voyages closer to Britain: thick wool undergarments and canvas jackets and trousers. Anyone who has been exposed to cold weather will know that base layers should be airy and light, preferably several thin layers to hold in the warmth. The combination of air and fabric provides insulation. Thick, woven wool on its own does not insulate so well. It is also harder to dry, and Parry's expedition could definitely expect to experience rain and sleet in the Arctic summer.

Parry writes that the men followed the same routine every day. They were woken at the same time, crept out of their sleeping bags that were lined with racoon fur, then they prayed together, before putting on their socks and boots that were either still soaking wet from the day before, or frozen stiff. Parry saw no point in drying boots and clothes when they would only get wet again. He was wrong. If you are going to succeed in the Arctic, you have to try to keep yourself as dry as possible. If you are constantly wet, it is more tiring and far easier to get ill. And this proved to be a major problem for the expedition.

One reason for Parry's optimism about reaching the North Pole was a new invention. The so-called sledge-boats were a combination of a boat and a sledge, with steel runners. Parry had studied the Inuits sleds and boats, and had himself helped to construct the sledge-boats. They were made from flexible wood, so it would be easier to cross the ice without any damage. This was a good idea, as anything that does not yield to the ice will be easily damaged. The plan was that when *Hecla* was stopped by pack ice near Svalbard, two sledge-boats would then be pulled north until they reached ice-free water from where they could then sail to the North Pole. The two-boat strategy applied to the sledge-boats as well. The two sledge-boats were called *Enterprise* and *Endeavour*, names that reflect Parry's optimism.

The expedition also had another new invention: a structure that was five feet in diameter, with four wheels, and could be put under the sledge-boats when the ice was hard. It also meant the boats could be rolled on over any unknown islands they might come across. The sledge-boats were twenty feet long, seven wide, with a flat bottom, and weighed 1,540 pounds, without freight. When they were fully loaded, each weighed 3,750 pounds. Each boat also had a nineteen-foot bamboo mast, fourteen paddles, and a tiller. Parry's plan was that the two sledge-boats would be pulled not only by the men, but also eight reindeer, and they would cover .6 miles a day. The calculated distance was not based on likely progress, but rather on what Parry hoped for, divided by the number of days for which they had provisions.

The reindeer were bought in Hammerfest, Norway, on the way north, along with enough moss for a daily ration of 4.4 pounds per reindeer. Parry explains in his book that reindeer were used as transport for the locals in Lapland, and the animals were obedient, which reinforced his belief that he had made the right choice. "Nothing can be more beautiful than the training of Lapland reindeer."

Later that summer, at around 80° N, Parry decided the time had come to try out the reindeer as draft animals. The *Enterprise* and the *Endeavour* were mounted on the wheels and the reindeer were harnessed to the sledge-boats. The crew and officers gathered on the ice to watch. Everything was ready for the push north, but the two heavily laden sledge-boats would not budge. The crew tried to pull with the reindeer, but the sledge-boats stayed put. The wheels had sunk into the snow and the steel caused friction with the ice, so Parry had to give up on both the wheels and the reindeer.

When preparing for the expedition, Parry had ignored the fact that they were likely to be traveling against the current for the most part, and in his book he writes how discouraging it feels to be drifting away from your goal. I know the feeling: As you take three steps forward, you drift one or two back. It is easy to sympathize with his desperation, but Inuits and whalers could have told him about the currents and that wheels would be of no use. Some days they progressed little more than .6 miles north.

Parry does not say so explicitly in his account, but it seems that he gave up all hope of getting to the North Pole fairly early on. Instead, the goal became managing to get farther north than Scoresby. Some days he writes about the sun, but more often than not there is mist, sleet, snow, and rain for days on end. The warmest temperature recorded was plus 64°F. On July 7, he writes that the men got "that wildness in their looks which usually accompanies excessive fatigue." According to Parry, the crew were still willing to follow orders, but they no longer understood the content of his orders. From July 22 to 26, they managed to move no more than 1.2 miles north.

Parry stopped and turned at 82° 45′ N, a quarter-degree farther north than Scoresby. The explorers were so exhausted that they forgot to hoist the British flag. The record held for forty-nine years. On their way back, they shot a polar bear and everyone who ate the meat got sick. Trichinella worms are often found in polar bear meat, but the bears are not born with the parasite—they get it from eating other polar bears. The meat should therefore not be eaten unless it is well cooked, fried, or frozen.

They were out on the ice for a total of sixty-one days and covered a distance of 172 miles. Parry reckoned that they had actually covered three to five times more, as they had gone back and forth so many times to help each other with the sledge-boats in their fight against the current. According to Parry, the distance was in fact "nearly sufficient" or equivalent, to walking to the North Pole in a direct line. Calculations like this can be comforting, of course, but have little other value. I can perfectly understand why Parry wanted people to know how much they had been through, perhaps as consolation for never making it to the Pole.

Parry does not write much about their food or weight loss, but I assume the men got thinner by the day. He lists the daily rations at the start of the book: .6 pounds hardtack, made from a mixture of flour and water, and .2 ounces pemmican (dried meat mixed with lard), as well as 1 ounce cocoa powder that they could mix with warm water. It is not easy to know how many calories this gave: The pemmican might have contained 1,500 calories, the hardtack 1,200 calories, and the cocoa 100 calories.[29]

Børge and I ate two pounds of food every day, which was mixed with water to bulk it out, and it gave us a total of 5,850 calories. The hungrier we got, the more water we mixed with the food to get that full, sated feeling. Thanks to all the research on food since Parry's day, we knew the required daily intake of proteins, carbohydrates, and fat. The aim was to prevent the body from breaking down faster than it needed to. We took a chance. If 65 percent of the total calories came from fat, we could save on weight, as fat gives twice as much energy as carbs. Physiologists generally maintain that the body cannot absorb food with such a high fat content in the course of a day. Fortunately, we found two physiologists who disagreed with this, and together we made our plan.[30] The challenge was that the body is not usually able to absorb so much fat, so in preparation for the expedition, we trained our bodies to absorb the amount needed and not defecate it out again immediately. I had fat in my porridge in the morning, drank hot chocolate made with cream for breakfast, lunch, and dinner, and drank olive oil with an evening snack. Thus we ate our way into the diet while we still had toilets nearby, and slowly but surely, the body got used to absorbing all that we ate and drank.

As we traveled north, the food was the same each day. Breakfast consisted of 3.2 oz. oats, .9 oz. formula milk, 1.8 oz. nuts, and .9 oz. soya fat. For lunch, we had 3.5 oz. oats, .9 oz. formula milk, .9 oz. raisins, .9 oz. sugar, and 3.5 oz. soya fat. And for dinner, we had 3.5 oz. dried meat, 3.5 oz. mashed potatoes, and 1.8 oz. soya fat. Everything was mixed with water, which we made by melting ice. Our reason for choosing formula milk is that it has more energy per ounce than other powdered milk.

We also had .4 pounds chocolate every day, which was specially made with extra fat—720 calories per 3.5 ounces—so it did not taste much like chocolate. Pure fat has 900 calories per 3.5 ounces. The chocolate was as hard as ice and had to be defrosted in our mouths, so every day we had to decide whether to close our mouths, and breathe through the nose, or keep our mouths open. The disadvantage of only breathing through your nose is that it is easy to get frostbite; the disadvantage of keeping your mouth open was that the chocolate did not get any softer. Not generally a healthy menu, but for getting to the North Pole, it did the job.

I initially thought the monotonous food tasted just fine, but as the days and weeks wore on, everything tasted like gourmet food. Long before we reached the North Pole, the food was the best I had ever eaten. To ensure that neither of us got more than our fair share, one of us divided the food into portions, and the other would choose which one he wanted.[31]

THERE IS AN illustration of Parry meeting a group of Inuits in north Greenland on August 8, 1818, when he was second-in-command on an expedition to find the Northwest Passage. The Inuits were part of the world's most northerly population, and the British did not know it was possible to live so far north. When they spotted people on land, they assumed that they were men who had been shipwrecked and sailed toward them. The Inuits, naturally enough, took fright when they suddenly saw two three-masted warships and retreated inland.

The drawing was a small sensation, as it was done by the Greenlander John Sacheuse, who was present that day. Sacheuse was an Inuit who had grown up farther south on Greenland, and in 1816 had been chosen to go with a whaler to Edinburgh. Here he learned to speak English and studied with an artist. The expedition had taken him on as interpreter, a smart move by the British team. The language spoken by the Inuits they met was different from his own, but they managed to understand each other and he was able to convince the Inuits that the Englishmen had peaceful intentions, which was true. The Inuits were skeptical to begin with. Sacheuse offered them gifts such as beads and clothes, but they were not impressed. When he then offered them a knife, they accepted it and agreed to meet the expedition members.

In the drawing, two officers are standing on the ice, dressed in the navy's blue parade uniform with gold buttons and epaulets. One is the leader of the expedition, John Ross, and the other is the next in command, Parry. Both are wearing tall, bicorne hats with a tassel at each end, and have a saber at their hip. And they both have the same type of narrow, black leather boots on their feet that they probably wore in the Admiralty offices in London.

The Inuits are dressed for Arctic conditions, with fur-lined hoods and sealskin boots, known as kamiks. They also have two light sleds, with four dogs harnessed to each, which should have inspired prospective polar explorers.

It was not only the British who were meeting people who looked different for the first time; the Inuits had never seen white people before, or such large boats before. They knew that people of similar appearance had been there before, from the stories passed down, but no one alive at the time had ever seen a European. It is difficult to know what the Inuits must have thought after the shock of seeing warships sailing toward them.

However, the drawing does give some indication. Two Inuits have been given a second knife and are holding up their gift with delight. The man in the top hat has given two other Inuits a mirror and they appear to be laughing at their reflection. A fifth Inuit is sitting on one of the dogsleds in the background, and looks to be laughing as well.

What no one could know is that the grandchildren and great-grandchildren of the Inuits living in the area would be crucial to future expeditions to the North Pole.

Thanks to the Inuit translator Sacheuse, the Inuits and the British were able to communicate verbally, and the expedition members stayed there for a few days before sailing on. Much to their surprise, the Inuits had iron. The polar explorers spoke about this on their return, and several later explorers searched for this source of iron, which proved to be three meteorites, found by Robert Peary seventy-six years later. He knew that the Inuits had hacked metal from them for generations to make knife blades, harpoons, and spearheads, and that they were dependent on the iron. But as soon as Peary found the meteorites, with the help of an Inuit, he scratched a "P" on one of them to show that from that day forth it was his property.

The British exchanged experiences with the Inuits. It seems that the Inuits were in turn invited on board the ships, and the expedition members were invited ashore. John Ross used his days well. He proved to be a good observer, and quick to learn. To his astonishment, the Inuits had no knowledge beyond their own area. It was no doubt a positive and educational experience for the British. It is harder to know what the Inuits felt. Even though they had access to iron, the iron the British had was of a completely different quality, thanks to the blacksmith on board. So every iron tool the Inuits were given was precious. And as they lived in a place without trees, wooden gifts may also have been welcome.

Apparently the Inuits thought that as there were only men on the first ship, there would be women on the other. A nice idea, perhaps, but that was not the way of expeditions at the time. Only men were on board.[32]

Some polar explorers tended to talk about the Inuits as children, as they knew and understood so little. In the time since 1818, however, history has proved the opposite to be true. The polar explorers were often as inexperienced as children when blundering into the Arctic. It was the Inuits who taught and helped them.

IN THE 1820S, Parry was criticized for not having used the opportunity to learn about the appropriate clothing, dogs, and nutrition from the Inuits he met on his expeditions. The criticism may well be justified, but to be fair, he did learn from the experience of the Inuits. He followed their tradition and Scoresby's recommendation to make any sledges flexible so they could cope with the ice, but unlike the indigenous people, he did not make them light enough. In addition, Parry knew only too well that footwear is every explorer's more difficult challenge, and he got the Inuits on Greenland to make sealskin boots for himself and the crew.

We also used kamik boots made by Canadian Inuits when we went to the North Pole. The sealskin boots were soft, comfortable, and nearly as high as the knee, which gave good insulation. Because sealskin is supple and the boots do not have hard soles, kamik boots are flexible enough to give your feet a firm grip on the ice, which was a real plus when we had to take off our skis to tackle hummocks. They worked very well for us—and no doubt would have done the same for Parry, had he not been a couple of months too late in setting out on the pack ice. It was summer and the ice had started to melt under their feet. In his book, he describes how they waded through water that had puddled on the ice and sometimes sank to their knees in slush. Kamik boots are almost waterproof, but no boots can keep your feet completely dry in the weather conditions in the Arctic.[33]

A New Kind of Writing

Parry wrote a book about each of his four last expeditions. The first three are about attempts to find the Northwest Passage, and the fourth is about the North Pole. Parry's books helped to establish a new genre in literary history, where men write about themselves and other men's feats in a matter-of-fact way, without sharing much of the emotion that attended those feats. The author and his men freeze, starve, fall ill, and die, or almost die, and boats are nearly wrecked—all described in a manner that makes heroes of the expedition leader and certain select men. The whole genre reeks of romantic idealism to me: men pitted against nature, facing extremely dangerous situations with courage and ingenuity, and without complaint; they win over every challenge and come through victorious. It makes for a very compelling narrative arc.

In the history of the North Pole, from Willem Barentsz to the present day, polar explorers have become more famous for their accounts of suffering and record-breaking than for their scientific work. The objective to carry out research was worthy, but did not excite as much interest in the general public. However, it was not enough that the men battled with nature on the ice and faced challenges that the reader could identify with. A travel account from the Arctic needed heroics: death, tragedy, survival, and dangers that surpass all others. The age of Romanticism may be over, but the narrative style persists.

The most widely read and discussed story about Arctic exploration in the first half of the nineteenth century is that of John Franklin's final expedition, where all the crew perished. It became a source of inspiration for future explorers.

Everything on board Franklin's ships, *Terror* and *Erebus*, was organized so that the officers would be comfortable while they searched for the Northwest Passage. The library held 1,000 books, the officers ate with silver cutlery, and they drank from cut-crystal glasses, and each ship was equipped with an automatic organ that could play more than fifty pieces of music.

When it did not return, the *New York Times* compared Franklin's expedition with the Crusaders' capture of Jerusalem and the Greeks' victory over Troy. Franklin was "ablest, oldest and bravest," and the newspaper used expressions like "religious heroism," "courageous endeavour," and "devotion to duty" to describe him.[34]

What was needed to become such a great polar hero? The *Times* wrote about Franklin's expedition in a way that echoed Edmund Burke: a goal should never be simple to achieve. This is a fair observation, but by preparing well an explorer should make the expedition no more dangerous than necessary. Parry had also experienced and survived many dangers, and was a very good storyteller. He remained a hero until he died. Yet he did not die with his boots on, but in bed, under a duvet, which did not excite the media or general public. Books that are published about Franklin still sell well, whereas Parry is now almost forgotten. If Franklin's fourth expedition had succeeded in sailing along the Northwest Passage, and his fate had not involved scurvy, tuberculosis, hunger, frostbite, death, and possibly even cannibalism, I doubt that it would have inspired so many poets and polar explorers, or been compared with the Crusades. Nor is it likely that there would be a marble bust of him in Westminster Abbey. The bust bears a legend written by the English poet Alfred Lord Tennyson: "Not here! The white north hath thy bones, and thou, heroic sailor-soul, art passing on thine happier voyage now toward no earthly pole."

I appreciate the romantic sentiment that Franklin is at the celestial North Pole but the Tennyson quote on the memorial cross for Captain Scott in Antarctica is more accurate in capturing a polar expedition. It contains both confidence and doubt, two feelings that most polar explorers, not only Scott, would recognize: "To strive, to seek, to find, and not to yield." The words are also a reminder that reaching the polar point or turning back before you get there can require as much courage. But also that the outcome of the expedition—death and glory—might well be the same.

Roald Amundsen, the person who eventually discovered the Northwest Passage, wrote that his childhood hero and stories about him "thrilled me as nothing I had ever read before . . . the thing . . . that appealed to me most strongly was the sufferings he and his men endured. A strange ambition burned within me to endure those same sufferings . . . I irretrievably decided to be an Arctic explorer."[35] Roald Amundsen never became a hero like Franklin; he described the first expedition to the South Pole in 1911 as one long ski trip, without any real drama. His book was very popular and his lectures about the expedition often sold out, but being the first person to return from Antarctica with all his men alive was not enough.

Most readers preferred the diary of the second expedition to reach the South Pole, written by Captain Robert Falcon Scott. It immortalized Scott through the injuries, hunger, and death they suffered. He did not write the published version of his diary alone. Without it being mentioned in the book, the manuscript was revised by J. M. Barrie, the author of *Peter Pan*, after Scott's death. The published version was far better written than Amundsen's and made Scott the perfect hero. He died in the correct way, in a tent, eleven miles from a food depot that could have saved him. His diary is both fascinating and touching, and his final words are: "It seems a pity. But I do not think I can write more."[36]

BOOKS ILLUSTRATED WITH photos became the preferred medium for explorers who wanted to tell about their expeditions.

When we went to the North Pole, I did not take my camera. One reason being that I was tired of people saying I would not attempt it if there was no interest in the media, and I wanted to prove them wrong. Another reason was that I doubted that photographs could truly capture the reality of the ice. But then Geir was injured. He and Børge had originally been the designated photographers, so I had to step in and take the camera.

I later came to appreciate that an expedition that is not documented in a way that is accessible to others really only exists in your head.

This attitude reflects George Berkeley's philosophy of knowledge, which is possibly more relevant now than ever. An expedition only exists in the way it is perceived. The best-known example of Berkeley's idea is a question he never asked himself: If a tree falls in a forest and no one is around to hear it, does it make a sound? For Berkeley, if only one person had experienced the event, that was enough for it to have happened. Yet one person is not enough in polar history. The legitimacy of expeditions was dependent not only on those having the experience, but also on it being shared with the greater part of a nation. For those of us who have grown up with photographs of the full earth taken by astronauts in space on December 7, 1972, it is hard to imagine a life without that perspective on our planet. If the cameras the astronauts were using had not worked well in space, I believe that the lives of many of us would be poorer for it, in terms of beauty, wonder, and humility.

But now, when it sometimes seems that the most important thing is to get the best photographs and distribute them as you go, I feel that my misgivings—that photographs rarely capture more than a fraction of reality—are stronger. For every photograph you take, you are less present in the moment. You become an observer of your own expedition, build an externalized relationship to the reality around you, and instead of leaving daily life behind, your thoughts remain in the world you have left.

On a demanding expedition, you *become* your actions. What you feel and think is not so important. For me, one of the greatest experiences on any expedition is the point when the past and future mean nothing, and I stop thinking and am completely present in the moment. It is this state of being that most polar explorers miss when they are safely at home.

It is only when my thoughts evaporate that things change for a few seconds, minutes, or hours. Only then do I experience the clouds, the sun, the stars, the ice, maybe some open channels, the wind on my face, the gentle melody of the frozen ocean. At some point in the future, and it is unlikely to be me, someone will be the first to go to the North Pole without a camera. The purpose will simply be to explore, not to be perceived as a polar explorer by others. And we will never know about it.

"The Land Which Is Not"

I am fascinated by how little the members of Parry's expedition knew about what they might find or how they might succeed, and yet were willing to risk everything.

Even when they were on the Arctic Ocean, all they knew about the world around them was what they could see if they climbed to the top of the ship's mast or stood on the top of an iceberg. Everything to the north, east, and west of the horizon was unknown. The history of exploration is, as the Norwegian polar explorer and humanist Fridtjof Nansen wrote, "a great unfolding of the power of the unknown on the human mind."

The first two lines of a Swedish poem, "The Land Which Is Not," by Edith Södergran, reminds me of how thrilling an unknown destination can be, and how it can awaken wonder in explorers, young and old:

> I long for the land which is not,
> For all that is, I am weary of wanting.

Nansen said that nowhere on earth have we progressed as slowly as in the Arctic, where every step has cost so much sweat, sorrow, and suffering. I agree with him, if by *we* he means Europeans and Americans, but whatever the case it is always hard to compare and quantify sweat, sorrow, and suffering. As Nansen points out, while all explorers experience suffering, nowhere has exploration promised as little material gain as the North Pole. And yet there were constantly new forces eager to push the world's northern boundary. For who, Nansen asks, is able to describe the feeling when "the last difficult ice-floe has been passed, and the sea lies open before him, leading to new realms"?[37]

WHEN I READ Parry's and other polar explorers' accounts of their expeditions to the North Pole, and what they achieved against all odds—the summer ice, the food, and in Parry's case the weight of the sledge-boats—I feel a great respect. They were, after all, children of their time, they fought as best they could, willingly risked their lives, endured great suffering, and achieved as much as they perhaps could within the system of which they were a part.

But one of the most striking things, for me, is what Parry and most other polar explorers from this part of the nineteenth century seldom write about: the sound of the ice. It is an experience like no other. It sings its own song. When the wind is not blowing, it can be still, with only the muffled, extended cracks and creaks of the ice that is constantly moving under your feet. The sound of the ice is like an unceasing melody. I have no idea if this is true or not, but it is as if the frozen ocean carries the sounds, that they are not carried by the air, but follow the ice. Ice floes bumping into more ice, pressing so hard that the ice is twisted up and slowly crushed. Winds from the south, east, west, and north, varying currents, and the swell of the ocean under the ice mean that the surface you find yourself on is never still. The ice breaks into new floes, channels open only to freeze over again, and new hummocks appear. The ice creaks and shrieks, grumbles and groans. The sounds meld. The South African explorer Mike Horn thought for some time when I asked him how he would describe the sound of ice, before he replied that it was like someone grinding their teeth in your ear. But unlike when someone grinds their teeth beside you, it can be difficult to discern which direction the noise is coming from, and whether it is far away or close by.

One night, Børge and I were woken by the noise of the ice around us breaking up, ice floes crashed into one another, creating new hummocks. It sounded like we were lying next to a stone crusher. We had set up camp on ice that we guessed was around two years old; younger ice is thinner and so not safe enough to sleep on. It was dark inside the tent and equally dark outside. We got all our equipment ready so we could leave the tent quickly if the ice beneath us, which was six to ten feet thick, started to break up. Then we lay there listening to the ice cracking around us. Ice being pulverized by more ice. At irregular intervals the noise and shaking would suddenly stop, and the night would fall silent, only then to start again after a few seconds or minutes.

When it got light in the morning, the ice, as far as we could see, in every direction, had been transformed. There was a wide, open channel only several feet from the tent. When the same thing happened during the day on another trip to the Arctic Ocean, Børge wrote in his book on the expedition: "Is that hummock over there not moving? Suddenly we notice that the whole horizon is moving! Soon everything is twisting—fast and furious . . . The entire area is being destroyed. To be caught on the ice like this with no possibility of escape is perilous." His companion, the Swiss polar explorer Thomas Müller, summed up the hardship of the ice, open-water channels, and hummocks that were appearing around them in a few words: "It's more and more of a mystery to me that Nansen and Johansen didn't shoot themselves."

But it is also precisely these characteristics of ice that make polar explorers want to go back. Ann Bancroft, the first woman to ski to the North Pole, in 1986, told me that the Arctic Ocean has its own magic and "alluring quality. It's hard, unpredictable, scary, pushes and demands of me in all ways, and the discoveries are inside and out." It is the sounds, the colors, the challenge. "The changing ice is both despairing and fascinating, which lures you back."[38]

BEFORE I STARTED going on expeditions myself, I sometimes smiled at the illusions people like Parry had about the North Pole, and how simple he thought it would be to get there. Now I shake my head at my old attitude. No one of sound mind would have traveled north. The history of expeditions demonstrates that human evolution "has always been carried forward by great *illusions*," as Nansen wrote. Because even though Columbus never discovered the sea route to Asia, without realizing it he rediscovered America instead. The voyage shows that being obsessed with an idea can change the world. The first expedition to spend a winter in Antarctica had not intended to do so. The people who have traveled north and south created dreams and inspired others to travel farther and farther beyond the horizon.

4

THE RACE TO REACH THE NORTH POLE

A political cartoon from 1882 illustrates the human toll of the era's rush to the North Pole, stoked by press attention and a desire for scientific progress.

Arctic Fever and Cannibalism

Interest in Arctic expeditions and the North Pole exploded in the 1840s and 1850s. In Great Britain, but also in the US, interest was fueled by the rescue operations to find John Franklin and the other survivors, and later the race between nations to reach the North Pole. It was Lady Jane Franklin, wife of the missing Franklin, who triggered what was called at the time *Arctic fever*.

The North Pole has always been associated with competition. Initially it was simply the race to get there first. But many races have since followed. I have only taken part in one, and that was to be the first to reach the Pole without the help of dogs, depots, and snowmobiles. Others have competed to be the first person to go there alone, the first person from their country, the first to go there and come back unsupported, the first to cross the Arctic Ocean in winter. When we went there, there were other expeditions, from Russia, South Korea, the UK, and Canada, who were also trying to reach the North Pole. It was a busy time, out on the ice. The British team were our toughest competitors. The expedition was led by the former SAS soldier Sir Ranulph Fiennes, who has been named as the World's Greatest Explorer in the *Guinness Book of Records*. Ran, as we call him, set the record for the expedition to get farthest north unsupported in 1986. The 1990 expedition was made up of two men, the other being the English doctor Mike Stroud. (Børge and I never understood why there needed to be a leader for two men.)

I was not aware that there was a new race to the North Pole until 1987. That year, I read an article in *National Geographic* magazine, the publication read by most explorers at the time, in which, having reached the North Pole on foot with aerial support, the Frenchman Jean-Louis

Étienne asked: "Will anyone ever reach the Pole under his own power, entirely unassisted?" For me, the question was an invitation to try.

THE MISSION OF Lady Jane Franklin's husband had been to explore the Arctic; her mission became to prompt, persuade, and pressure the Admiralty and other private institutions to search for him. Between 1848 and 1859, there was a race to find any survivors. More than fifty expeditions set out from the US and Britain. The search team captured Arctic foxes and attached written messages to them addressed to Franklin and his men, giving the position of rescue ships. Then the animals were released again. The hope was that Franklin might shoot the foxes for food and so find out that a rescue operation was in the vicinity. A similar message was attached to balloons, which were then released to float over the ice, water, and skerries. Private and government expeditions competed to find clues as to what might have happened. There is no record of how many people died during these rescue attempts, but having read numerous accounts of expeditions to the Arctic and subsequent rescue operations, I think that far more people died in connection with the rescue than with the expedition itself.

Six years after the race to find Franklin had begun, the explorer and surgeon John Rae, from the Orkneys, returned to London in 1854, following one such rescue expedition. He reported that the Inuits had noticed that the British refused to eat the blubber and innards from seals and walrus so essential to the Inuits' diet, and that they often died from scurvy and starvation as a result. In his report to the Admiralty, Rae described the remains of some bones he had found, presumably from Franklin's expedition, that looked as if they had been cooked and cut with a metal knife, and from "the contents of the kettles, it is evident that our wretched countrymen had been driven to the last dread resource—cannibalism." It seems that the men of the Admiralty had grown tired of sending new expeditions north, and wanted an end to it, so they leaked this information to *The Times*. People were appalled. The idea that civilized men could eat each other was horrifying to the Victorians.

Readers of newspapers, books, and periodicals, and lecture audiences, had lapped up the details of countless rescue operations and the uncertainty surrounding the fate of the polar explorers. In the spirit of Edmund Burke, they were both captivated and repelled by the terrible sequence of events. But, as Burke had predicted, the readers' experience of the sublime faded into terror when the specter of scurvy, starvation, and cannibalism loomed.

A good "sublime" story should stoke two strong emotions in the reader: first, shock or helplessness, then enormous relief at what rational people can achieve when faced with grave danger. The latter was missing from the story of Franklin's final expedition.

Franklin's wife, Jane, and the writer Charles Dickens were two of the most vocal critics of these claims of cannibalism. It must be understandably hard to accept that your husband may have eaten one of his own crew. Dickens's explanation of why he thought it was not possible is more interesting—and more revealing. He claimed in his own periodical, *Household Words*, that the English, with their superior culture, could never be tempted by what he, quite correctly, called "the last resort." It must have been the more primitive creatures, the Inuits, whom he preferred to call "lying savages," who had succumbed to cannibalism. Dickens did not accuse Rae of lying, but said that the stories the Inuits had told him were nothing more than "the chatter of a gross handful of uncivilised people" and should be left untold. For Dickens, it was natural to think that a good

Englishman would rather starve to death. Unlike other nations and races, they were far too civilized to eat human flesh.[1]

A News Revolution and the American Route

The interest in expeditions to the North Pole through the nineteenth century to 1909, when Robert Peary and Frederick Cook claimed to have reached the Pole, must be seen in light of the technological developments in the same period. As interest in the North Pole and polar explorers like William Parry and John Franklin grew, the first steps in the globalization of news and literature propelled their stories around the world. The culture of the international celebrity was born, and polar explorers became famous throughout the greater part of Europe and eventually the US.

Until the start of the 1800s, access to literature, be that novels, poetry, or nonfiction, was limited. In the first half of the century, thanks to new technology, literature started to be mass-produced, and could be distributed more efficiently by train. More people learned how to read, gas lighting became more readily available in people's homes for them to read by, prices also became more affordable as printing became steam-powered, and as living standards rose, people gained more free time and disposable income. The novel *Frankenstein* and books by William Parry and John Franklin were sold in bookshops and kiosks, often in train stations.

Telegraph lines were installed along the railways, and suddenly news could reach the towns and cities in no time at all. The expeditions and their heroic feats became headline news. Newspapers published national and international news every day. In August 1858, a telegraph cable was pulled along the seabed between the US and Great Britain. It fell apart three weeks later, and a new cable was laid in 1865, which also broke. Another cable was successfully laid the following year and cross-Atlantic communication shrank from two weeks to two minutes.[2]

Expeditions and newspapers were heavily dependent on each other. The media needed a constant supply of new heroes and dramatic stories. And the more prestige an expedition could muster, the more money and state support followed. Newspaper barons planned and organized their own polar expeditions, and the rights to stories and books were sold before the expeditions had even set off. Sometimes journalists even became explorers themselves, as was the case with Henry Stanley, who went off in 1871 to find David Livingstone in present-day Tanzania.

THE AMERICAN POLAR explorer Anthony Fiala once said that one does not give up until "the command given to Adam in the beginning—the command to subdue the earth—has been obeyed."[3] It seems that this became a kind of creed for the US, which was starting to claim its position as a polar nation.

The expeditions were militant in their ambitions to conquer nature, map the ocean, carry out scientific studies, and acquire territories on sea, land, and ice. They were also militant in their treatment of other people and cultures, and assumed it was their duty to tame and subjugate the forces of nature. Polar history entered an era that Joseph Conrad described as *Geography militant.*

Like several other polar explorers, the American Elisha Kent Kane (1820–57) was inspired to travel north when he read about the missing John Franklin. Kane lived in Philadelphia and was involved in the latter part of the US's invasion of Mexico in 1848 and 1849. He had no experience of the Arctic, but was convinced that America should help to find Franklin. And not only should America offer its assistance, but he himself was the best person for the job. Kane took part in two rescue operations before he put together his own expedition in 1853. It was far smaller than the British expeditions, but he set off north as captain of the brig USS *Advance.*

As soon as there was the slightest bit of swell, Kane was violently seasick. He also had no idea how to navigate, had no experience of leadership, and was plagued by rheumatic fever. Only one of the men on board the *Advance* had any relevant experience from the Arctic region. The reason the crew was so unsuitable was that it had been hard to find volunteers. Kane had eventually been forced to take on sailors without work who were hanging around the harbor.

Instead of heading toward the area where Franklin might be, Kane decided to look for him somewhere he was highly unlikely to be: Smith Sound, which lies east of Ellesmere Island, closer to Greenland. His intention was in fact to find a new route to the North Pole, fighting his way through to ice-free waters, so as to be the first to reach the Pole. By saying it was a rescue operation, he cleverly secured financial support and a boat. American sponsors liked the idea that their young nation could take over where Great Britain had failed, and save the superpower's missing superhero.

When Kane's expedition reached northwest Greenland, he proved that while he might be a poor sailor and inexperienced polar expedition leader, he was a sound pioneer. He brought something new to polar exploration. Unlike Parry and Franklin, he spent time with the Inuits to learn how to drive a dogsled, the best mode of transport for crossing the ice.

In 1854, Kane's plan was to abandon the *Advance* when she became beset by ice, and carry on north. But he made the same mistake as countless polar explorers. He gave in to his restlessness and set off too early, when the temperature was around –40°F. He ended up having a mental breakdown along the way. This seems only natural to me, given that he had started the preparations to become an explorer only some years before, by spending a few nights in a tent close to his home city of Cincinnati.

As they trekked north, Kane became convinced that a member of the expedition was a polar bear and gave orders for him to be shot. Fortunately, no one obeyed the order. But two of the crew died later on their return: one from tetanus and the other after his foot was amputated. Kane did not let the deaths or his own mental health deter him. He headed north again the same summer, with those who were still healthy enough to go. At some point, one of the team climbed a ridge to look around.[4] To the northwest, he saw open water and clouds heavy with rain, and believed that the expedition had discovered the ice-free Arctic Ocean. Pierre Berton concluded in his book *Arctic Grail* that it was a combination of wind, waves, and wishful thinking.[5] It is hard to know how convinced the man was of his discovery, but Kane, who was waiting down by the sleds, was in no doubt. His expedition had achieved what generations of explorers had dreamed of: open water all the way to the North Pole and the Pacific. The sensational discovery would justify the expedition, but still he felt they could achieve more. Just like myself, Kane also wanted to impress his father. His hope was to be able to "advance myself in my father's eyes by a book on glaciers and glacial geology."[6]

The intention was to sail home again in late summer 1854, but the ice did not melt as they had hoped and *Advance* remained trapped. Kane decided that they would all stay with the boat for a second winter. They had managed to keep the temperature on board the boat above zero the winter before thanks to the Inuits who had taught them how to insulate it with moss and peat. But the crew mutinied, and some of the men started to walk south in the hope of avoiding another winter.

Nearly a year later, on May 20, 1855, *Advance* was still stuck in the ice. Kane and the remaining men abandoned the boat and walked for eighty-three days to Upernavik, on the west coast of Greenland. In a farewell letter left on board, in case they too went missing and someone found the boat, Kane wrote that they had so little coal left that they would soon have to start burning the ship's timbers. Staying aboard was no longer an option. The last things they did on board were to pray and quietly pack a portrait of John Franklin.

Only one man died on the return journey—no mean feat, given all that they lacked and the number of men with scurvy. The US dispatched two ships to find and rescue the expedition in summer 1855. Two years of arduous struggle, frostbite, hunger, and very poor hygiene had rendered the men unrecognizable. Even Kane's brother, who took part in the rescue operation, failed to recognize his dirty, exhausted, bearded brother when they met.

THE RECEPTION BACK in New York was extraordinary. They were heroes. Kane said that a man who spends a year in the Arctic ages faster and more visibly than anywhere else. Fridtjof Nansen's dry remark, many years later, was that the brutality of Kane's experience was due to his hopeless preparations.[7] In part, this is true. The book that Kane wrote sold 65,000 copies; he portrayed himself as an exemplary leader and the world believed him.

Kane understood that a successful book would mean a good income and fame, and that it took more than matter-of-fact stories about what happened based on diaries and logbooks. A polar explorer had to describe how the most northerly record was won through inhuman suffering and heroic daring, for the benefit of the nation. As Berton points out, Kane's book was "a striking example of the power of the pen."

The small blue plaque on Kane's grave in Philadelphia, the city where he was born, says he was the first "to chart a course" toward the North Pole. A number of expeditions later followed this course along the west coast of Greenland, up through Smith Sound, Kane Basin, and north. It became known as the "American route" to the North Pole. The polar explorer's initial proposal to lead a rescue operation in search of John Franklin was long forgotten.

When he died in 1857, Kane's coffin was sent on a two-week cortege tour of America, as a final honor, from town to town, by train and steamship, before he was laid to rest. The funeral is said to have been the biggest in the history of America, until that of President Abraham Lincoln.[8]

Charles Francis Hall's Obsession

"Americans can do it" was the conclusion of the leader of the second American expedition to the North Pole.[9] Not only was Charles Francis Hall (1821–71) obsessed by the North Pole, but he was also of the mind that he was born to get there first. And once he had achieved that, there would be nothing left to live for. "I shall be perfectly willing to die," he wrote.[10]

President Ulysses S. Grant had also caught the polar fever and must have believed Hall. The very first time the aspiring polar explorer and the president met in the White House, they discussed which route Hall would take north, and decided he should follow the route that Kane had charted. Grant offered Hall one of his custom-made cigars, and they went on to discuss the challenges that awaited him. Congress later granted $50,000 and stated that the expedition was "under the authority and for the benefit of the United States."

Hall had worked as a blacksmith and engraver, before setting up his own newspaper. He had also taken part in two rescue operations to find John Franklin. I always recommend that anyone who wants to go to the North Pole read as much as possible about past polar expeditions, to learn about the ice, clothing, food, and equipment, as well as talking to people with actual experience of going there. It is important to be humble and learn from others, until you know what is best for you. Hall did just that: He read everything he could get his hands on, and when he went on the rescue operations, he was curious about the history and culture of the Inuits. He learned to build igloos, to hunt and drive a dogsled, and taught himself to like the slightly harsh taste of seal and blubber. According to Inuit tradition, blubber gives you strength when you are cold and tired, which is true in my experience. When you are exhausted and freezing and have been walking for two weeks and it feels like all the fat on your body has been spent, blubber tastes like the best thing on earth.

Hall was careful to point out that he was an idealist before he set off for the North Pole. He was not driven by paltry matters such as honor or money. He had only one motive: "My desire is to promote the welfare of mankind."[11]

When he was about to set off in USS *Polaris*—a steamship with sails—from New York on June 29, 1871, it became clear that his crew were not quite so high-minded. The British had always relied on naval men, but Hall had chosen whalers and ordinary sailors. The steward was so drunk that he could not board the ship, and the first chosen cook abandoned the trip with three other crew members before sailing. Hall was convinced that for the expedition to be successful, he needed to work closely with the Inuits, so he paid some men to accompany them north to hunt and some women to make food and equipment. The Inuits, their children, and the Americans lived together for long periods.

Later that September, the idealist Hall set out from Thank God Harbor, which he had named their camp, on a sled with an Inuit guide to break William Parry's record for most northerly point, but failed dismally. When he returned to *Polaris* two weeks later he gulped down a cup of coffee before even taking off his fur. A few minutes later, Hall complained of a headache and threw up. In the evening, he became paralyzed on the right side, and the scientific leader of the expedition, a German doctor called Emil Bessels (1847–88), said that in all likelihood Hall had had a stroke. Hall denied this and accused his men of poisoning him. He was particularly suspicious of Bessels, who treated him when he was sick. Hall died on November 8, 1871, after several weeks in great pain. Bessels concluded that the cause of death was another stroke. The doctor wrote about the preparations for his funeral: "While we were dressing the corpse the coffin was being hammered together in the engine room. The hammer blows reverberated sadly and hollowly through the noiseless silence of the arctic night." The coffin was to be buried six feet underground in the permafrost, but after two days the men gave up trying to dig any deeper. Hall was buried two feet underground. A snowstorm was raging during the funeral, so there was only enough time for a short speech and a prayer. Each of the men around the grave took a handful of frozen earth and scattered it over the coffin. Bessels's account ends with a quotation from Dante's *Divine Comedy*, Canto 32, Ninth Circle: "Each kept his face turned downward; from his mouth, the cold, and from his eyes, his saddened heart provides itself a witness in their midst."[12]

A very elegant choice, on first glance. But perhaps some of those at the graveside knew where the words came from, in which case I am sure a shudder ran down their spines. This circle of Dante's hell is reserved for people who have betrayed their country or family. It contains the mysterious river Cocytus, a frozen river of human tears.

In Dante's tale, the dead traitors are buried in ice in four different ways, depending on the gravity of their betrayal. They might have ice up to their middle, or neck, or be head down in the ice with their legs sticking up, or totally buried in ice, as Hall was. Of all possible literary references, Bessels chose this as Hall's final goodbye, and it is not likely to have been in jest or out of respect.

Hall's death was never investigated as suspicious. It would also have been complicated by the fact that several logbooks and diaries had disappeared, which in itself is suspicious. In his book about the expedition, *Die Amerikanische Nordpol-Expedition*, Bessels quotes from the missing material three times, without mentioning where he had read it.

Fortunately, Hall's grave was exhumed by his biographer, the American Chauncey C. Loomis, in 1968. Thanks to the permafrost, Hall's body and clothes, and the American flag he had been wrapped in, were all intact. The wood of the coffin—which was far taller than Hall, suggesting that it may have been made for someone else—looked fresh. Loomis took samples of his nails, bones, and hair; Hall was then wrapped up again as he had been found, the lid was put back on the coffin, and he was reburied in the same grave.[13]

The samples showed that Hall had died from arsenic poisoning. He had not only been poisoned once, but several times in the last weeks of his life. Every time Hall felt a little better, he was given another dose of arsenic, causing him to writhe in pain.

Two things indicate that Hall had been right in his accusations against Bessels. As ship's doctor, Bessels had access to arsenic, which was seen as a kind of panacea and used in small doses to treat arthritis, headaches, cancer, syphilis, and other illnesses. It was usual on expedition boats for the doctor to have control over the medicine cabinet. They were locked away to avoid any abuse and the doctor kept a record of what was used. In Bessels's book about the expedition, which was published in German in 1879, but not translated into English until 2016, he says nothing about any arsenic going missing or being stolen.[14] The expedition is described in detail, so it would be natural to mention if a deadly poison like arsenic disappeared.

Bessels had three possible motives for killing Hall. The first was a deep contempt for Hall that he shared with many others on board. I fully understand that contempt, or even hate, might drive someone to murder, especially if you are living in close quarters on board a ship in the Arctic for an indefinite time, from one season to the next, isolated from the rest of the world.

A particular feature of being so isolated and stuck is that it is difficult to avoid thinking the same thought over and over again. You cannot get away. I have several times tried to rid myself of a thought when isolated, but when there is nothing or no one to give you new impulses, both positive and negative thoughts are amplified. The thought becomes a kind of isolated madness, even when it is something far more innocent than possible murder. If you are hungry, you naturally think about food a lot of the time, but also have other irrational thoughts. Another motive for Bessels was that if he killed Hall, he could himself lead an attempt to reach the North Pole. According to Bruce Henderson's book *Fatal North*, Bessels later tried to bribe others on board to help him set the new record for reaching farthest north.[15]

The third motive that Bessels might have had, which no one else did, was jealousy. He was in love with the same woman as Hall. It is of course not unusual for two men to love the same woman, but it is perhaps more unusual for those two men to then go on an expedition to the North Pole together. The woman in question was the talented and beautiful American sculptor Vinnie Ream (1847–1914). When she was only seventeen years old, she was asked to make a bust of President Lincoln, and he was persuaded to sit for her for five months. The white marble bust still stands in the rotunda on Capitol Hill in Washington, DC. According to her biographer she was "not always able to establish long-lasting relationships" but was better at short relationships.[16]

Hall met Ream early in 1871 when he was in Washington, DC, making preparations for the expedition to the North Pole. She was attracted by his "bear-like quality." Hall and Ream had dinner together on a number of occasions, and Bessels was invited to join them every now and then. That was a mistake: "Hall enjoyed Vinnie's company, but Bessels became instantly infatuated with her." Bessels has been described as a short man with a thick German accent. Hall writes in his last letter to Ream before departing on June 28 that he thinks about her "all the time and anticipating the pleasure of seeing you tomorrow" but that they had to leave earlier than planned and were sailing the next day.

Hall was married but had been in the Arctic for nine years in the 1860s, and barely visited his wife when he was home. Unlike many other polar explorers, he did not engage in sexual relations with any Inuit women. Perhaps he wanted to be loyal to his wife, or was thinking of the potential children who would be left behind. He may even have been concerned about the health implications, and sexual diseases—as Western men transmitted their diseases to the women, and vice-versa.

When Hall's expedition arrived in Upernavik on Greenland, he received a last letter from Ream. The post boat sailed faster than the heavily laden *Polaris.* She also sent him a copy of the bust of Lincoln, having previously given him a picture of it. He had the bust in his cabin and hung the picture below it. In his final letter to her, he wrote: "You may expect that when you again hear from me and my company, that the North Pole has been discovered."[17]

The murder of Hall reminds me of an Agatha Christie novel in which someone is murdered and everyone who might have killed them is within the same four walls. And just as in her stories, it was possible to uncover the culprit—only in the thrillers is it beyond all reasonable doubt. Out on the ice, things were far murkier and less easy to resolve.

THE EXPEDITION CONTINUED on after Hall's funeral, and it seems that Bessels gave no more thought to Hall. In his book, he writes with respect about the Inuits they meet and their knowledge of geography. The Inuits had no traditions for drawing maps, but the best navigators memorized the landscape—the mountains, plains, headlands, lakes, and fjords—so whenever Bessels asked about the geography and routes, he was always given detailed replies. When he suggested they could draw the map they had in their heads on paper, they were not interested. They had never needed it.

Bessels writes that he got closer and closer to the North Pole in his dreams. He passed one latitude after another and discovered things he knew would shock the world. But it was only in his dreams.

In the early morning of October 15, 1872, *Polaris* collided with an iceberg. Several of those on board feared that the boat would sink and nineteen people managed to scramble onto an ice floe with food and equipment. *Polaris* did not sink, but the ice floe drifted away from the boat, so they were cut adrift.

After Hall's death, members of the expedition had continued to live with the Inuits on board the ship. Among the nineteen people stranded on the ice floe were five Inuit children, the youngest of which was a newborn baby. Emil Bessels was still on the ship. The adults on the ice floe suspected that he had seen them, but made no attempt to help them. They were stranded in one of the most inhospitable, windswept regions in the world, and would soon have to start rationing their food.

They drifted on the ice floe for nearly seven months at the coldest time of the year. They traveled a total 1,800 miles south, from Smith Sound down the coast of Greenland, past Baffin Island to the mouth of Hudson Bay. They were finally rescued by passing sealers at Labrador on April 30, 1873, shortly after a breakfast of seal intestines. The captain of the vessel asked: "How long have you been on the ice?"

"Since the fifteenth of October."

The captain could not believe what he was hearing: 196 days and nights on an ice floe, with no preparation. They had of course nearly drowned, frozen to death, and starved to death. One of the Inuit hunters also said that he had been afraid, with good reason, that he would be killed and eaten when he failed to catch anything. The temptation must have been there, when they were starving and there was nothing to eat. They had built themselves igloos and lived on seals, narwhals, birds, and polar bears, when the hunting was good. All nineteen survived.[18] I have skied and camped on Baffin Island and was well prepared, which makes it even more incredible that anyone could survive in that cold on an ice floe with minimal equipment.

When you are floating south to warmer climes on an ice floe, the only thing you can be certain of is that at some point the ice beneath your feet will melt.

The Imperial and Royal Austro-Hungarian North Pole Expedition

The majority of explorers wanting to reach the North Pole, from the mid-1800s up until the First World War, were American, but they were not the only ones to sail and walk toward to the Pole. The Imperial and Royal Austro-Hungarian North Pole Expedition started on June 13, 1872, when the American *Polaris* expedition was still missing. The expedition was unique in an Arctic context, in that it was multiethnic; all the men were from the Austro-Hungarian Empire, to be fair, but spoke German, Italian, Slavic, Hungarian, and other local languages from the Tyrol. Italian was the lingua franca. A number of the men could already ski and had experience of the snow and cold from the Tyrol. The ship, *Admiral Tegetthoff*, was a three-masted bark, with a reinforced hull to deal with the ice and a steam engine.

Four years prior to departure, Julius Payer (1841–1915), one of the expedition leaders, had never even thought of traveling north. That all changed when he came across a newspaper article about the North Pole when hiking in the Alps. In his own account of the article, Payer says that he was "filled with astonishment . . . that there should be men endued with such capacity to endure cold and darkness. No presentiment had I then that the very next year I should myself have joined an expedition to the North Pole."[19] That expedition barely got going, but the second expedition he took part in reached 77° 10′ N. And now he was going to make a third attempt.

This story shows how swiftly one can make the decision to risk one's life to reach the North Pole. It reminds me how absurd and irrational the impulse is, and that it's rarely the result of calm, reasoned consideration. I also recognize myself in the story. When I had the idea to be the first person to ski to the South Pole alone in 1992, I decided to do it immediately. It was going to be me! Then reason kicked in and I thought rationally about how I would manage it. Had it been the other way round, and I had been reasonable first, then made my decision, I probably would not have gotten there before anyone else.

Everyone on board the *Admiral Tegetthoff* had to sign a declaration renouncing "every claim to an expedition for our rescue, in case we should be unable to return." The expedition's sponsors in Vienna had obviously learned from the countless operations sent to rescue British and American expeditions, and were careful to avoid any liability. It is difficult to know precisely how many rescue operations have been sent to the Arctic. Whatever the case, there have certainly been more rescue operations than expeditions to the North Pole, and Northwest and Northeast Passages.

The Austrian author Christoph Ransmayr, who wrote the novel *The Terrors of Ice and Darkness* about the Austro-Hungarian North Pole expedition, mentions that the ship's hunter, Johann Haller, only used an exclamation mark twice in his diaries, in all his years on board.[20] Both were used on the day that the machinist—the only one on board who knew how to repair the steam engine—died. Haller must have been a levelheaded man. He was Payer's climbing friend, from the time they read about the North Pole when camping in the Alps. Toward the end of a letter that Payer sent to his friend Haller, urging him to join the expedition, he writes: "We will face cold and danger—does that frighten you?"

Ransmayr quotes from diaries written by members of the expedition, but the rest of the book about the days in the north are fiction. He describes life on an expedition to the North Pole better and more realistically than most participants, as he frees himself from a tradition where polar explorers' emotions and inner lives are scarcely mentioned.

The expedition was headed by two people: Carl Weyprecht (1838–81) was the leader on board the ship and on the ice, and Payer was to take the lead when they left *Admiral Tegetthoff* to study and explore the Arctic Ocean and find their way to the North Pole. In his own book, Payer says that he wants to complete the journey started by Pytheas some 2,100 years earlier, and surpass Alexander the Great in finding a quicker route to India: in other words, "*the route through the ice*—the most perverse notion that ever entered into the mind of man to conceive."[21]

August Petermann—The Belief in an Ice-Free Arctic Ocean

Payer named the most northerly island that he saw, but did not reach, after the geographer and cartographer August Petermann (1822–78). Petermann was a champion of new expeditions to the north. He was one of the most famous scientists of his day and perhaps the last person who tried to prove that the Arctic Ocean was ice-free. Like Ptolemy and Mercator, also cartographers, and Isaac Newton, who never left England, he preferred not to travel himself. His hope was that new expeditions to the North Pole would be able to prove his theories about the Arctic Ocean.

He thought it was naive to use the American route, to sail through Smith Sound and north, as explorers would then be obstructed by land to the north of what is now known as Canada. Petermann believed that Greenland stretched further north than it actually does, and continued past the west side of the North Pole. The obvious route would therefore be to follow the Gulf Stream, up past Norway and on into an ice-free Arctic Ocean.

Petermann's theory of an ice-free ocean inspired two German North Pole expeditions. These expeditions took place around the same time that the Iron Chancellor, Otto von Bismarck, had started the unification of Germany by winning wars against Denmark, Austria, and France. The first expedition in 1868 was comprised of only one ship, *Grönland*, which reached Svalbard before returning. The second expedition the following year had two ships, bearing the proud names *Germania* and *Hansa*. On the second expedition, the boots were apparently so awful that he decided to make his own. When a polar bear was killed, Payer removed the meat, sinews, and bones from the polar bear's legs and paws and used them as boots.

Germania and *Hansa* got separated—something to be avoided at all costs in the Arctic Ocean. *Germania* raised a flag somewhere along the east coast of Greenland to signal to *Hansa* that they should anchor together, so the captains could discuss and agree on the plan from there. But the captain on board *Hansa* thought the flag meant they should sail west, and he gave orders to change course. The *Hansa* was crushed by ice on October 22, 1869, but fortunately the crew managed to empty the boat of food and equipment and survived on an ice floe that drifted south. They also chopped down the three masts for firewood.[22] And in June 1870 they eventually came to Greenland. Neither *Hansa* nor *Germania* made it farther north than Greenland.

On his third expedition, Payer writes about the dark, cold, and loneliness, and that he misses home. His diaries are full of feelings and questions, and detailed accounts of arguments. The men on board try to communicate with the world by sending messages in bottles that they regularly drop into the open leads in the ice-covered ocean. The desire to use bottle post is understandable: no one knew if they would survive and the hope was that at least some of the bottles would reach civilization.

The plan was to stay for two winters. As time passed and they failed to find the ice-free ocean or new land, Payer writes that even a small cliff would boost their "self-esteem as explorers." They then discovered the archipelago on the northeastern coast of the North Pole ice that they named after the emperor, Franz Josef Land, and the island Petermann Island.

Walking on Thin Ice

Crossing thin ice is something you have to do every day when you are heading for the North Pole, so it is worth learning to walk with a light step, and you should preferably take skis. On skis, your weight is distributed over 6.5 feet, so you can cross far thinner ice than the men on the Austro-Hungarian expedition could manage. Even though several of them came from areas with plenty of snow in winter, no one had thought to bring skis.

The ice in the Arctic is different from ice elsewhere because it is saline; it is elastic and wobbles when you cross thin areas. Børge and I compared the sensation with walking over a waterbed. One day I was skiing ahead of Børge over some thin ice and there was a stiff northwesterly blowing, so it was hard to hear anything but the wind. Our conversation went as follows:

Wobbling? I shout.

Yes!

What?

Yes!

A lot?

Yes!

I checked the thickness of the ice by jabbing the tip of my ski pole into it at every step. A day or two earlier there had been open water where I was walking. When I heard Børge shouting *Yes!*, I lifted my arm again and hit the ice with my ski pole. The rule of thumb was simple: If the pole went through the third time you hit it, the ice was thick enough to walk on; if it went through the second time, it was risky; and if it went through the first time, it was extremely dangerous. I walked another few feet and then the pole went straight through the ice with no resistance. Suddenly it was seesawing more than before. I tried to turn. One foot, ski and all, disappeared down into the water. I was afraid that the weight of me on only one ski would be too much for the ice, and lay down on my stomach to distribute the weight of my body, then wriggled backward. I pushed the pulk in front of me first, then wriggled past it and started to pull.

A thought suddenly popped into my mind, as it sometimes does when I am in the north: It's a good thing my mother can't see me now—she's had enough to worry about as it is—and then it disappeared as quickly as it had come.

Once I had wriggled back to a place where the ice was more stable, I could stand up and follow Børge. He had started to walk in a wide semicircle around where I had gone through, and then turned north again, constantly checking the ice with his ski pole.

Never Return

There are two stories from the Payer expedition that I think are illustrative of the kind of attitude that he and his chosen men had. The first is from March 26, 1874, when they left *Admiral Tegetthoff* for the second time, to look for unknown lands in the north, and hopefully reach the North Pole.

From Payer's diaries, we know that his group of men had a headwind and it started to snow as soon as they set off. Just over a half mile from the ship, as the crow flies, visibility was so bad that they could not work out which way was north and which south. They started walking in circles. Payer concluded that it was impossible to continue and the most sensible thing would be to go back. "Nonetheless, we preferred to pitch the tent behind a hummock, where it was not visible from the ship," he writes. They set off again early the next day, in the hope that no one from the ship would spot them. After seventeen days, they had walked 186 miles and set the record for reaching farthest north at 82° 5′.

Perhaps Payer and his men thought the record was unbeatable? It is quite usual for polar explorers even today to think that the record they have set will not be broken by others. You suffer and freeze so much that you cannot believe that anyone else could endure the same sacrifice.

The second story is from when the expedition team are about to abandon *Admiral Tegetthoff* on May 20, 1874. The ship is frozen into the ice, and the only way to survive is to pull three Norwegian whaling boats south over the ice until they find open water. Because the boats are so heavy, all the men have to pull one boat first, then go back for the second, and then the third, so the distance they cover is five times more than they have progressed. When they reach the open sea, the plan is to sail or row until they find another whaling ship or another vessel.

They have to travel light, so everything that is not strictly necessary is left behind. Even pictures of family and loved ones. Instead of leaving the pictures on board the ship, the officers and crew decide to do them the final honor of nailing them neatly to the surrounding walls of ice, formed by the weather and currents of the Arctic Ocean. The ship is draped with two Austro-Hungarian flags from the top of the masts.

After ten hours, they have moved .6 miles south of the ship. After two months, they were nine miles from where they started.

Weyprecht, who was responsible for their retreat, kept a diary in which he says he is astounded that he is able to keep calm, "and sometimes it is as if I am not part of this at all."[23] I know the feeling; when I have been at my most exhausted, it feels as though my mind and body are separate entities. The pain remains physical, but it almost feels as though it belongs to someone else, who happens to have the same body as me.

ON THEIR RETURN to Vienna, a total of 834 days after they left, the survivors were welcomed like heroes. However, some leading figures in Vienna did question if Payer had really discovered new land, and if Franz Josef Land and Petermann Island actually existed. Payer left the geographical society in protest. It is easy to sympathize with him, given all that he had been through, and then not to be believed. Payer moved away from Vienna for a few decades and in that time proved to be a talented artist. His paintings sold well, though the larger paintings were often harder to sell, but he was stubborn and continued to make large painting after large painting even though no one wanted to buy them. His pictures are still being auctioned in Europe. They are extremely realistic, often with motifs from his own life—polar bear hunts, storms, men toiling, and the Arctic sun barely above the horizon. Payer also did a painting of John Franklin's last expedition; when you study the painting it is as if Payer is giving life to some of the pain he experienced in the north, when illustrating the Englishman's death in the frozen wastes.

His best painting is perhaps *Nie zurück* ("Never Return"), which shows the march south from *Admiral Tegetthoff.* It is almost dark, the men are kneeling, lying, sitting, and standing, some are awake, others dozing, the birds that will probably eat the first to die hover overhead. Several faces are illuminated by an unknown, almost religious source of light. Weyprecht stands alone in front of his men, who are turned toward him. He has the Bible in his right hand and his left arm reaches out toward the sun and distant lands. The motif reminds me of Jesus and his disciples, or Moses leading his people.

PAYER'S LAST GREAT dream, when he was seventy, was to reach the North Pole in a submarine. The idea appeals to me. Ransmayr writes about how Payer might achieve this: set off from Kiel, in northern Germany, then dive down under the surface of the ocean, and when he was directly below the North Pole he could send up an explosive charge to lift the lid of ice and blow it apart. The open, shining water will then reflect the sky, and then he can finally rise up out of this mirror.

British Arctic Expedition

"England is not too rich to be bold, too luxurious to be simple, too cynical to be pious, too genteel to believe in honor and glory and the sweetness of self-devotion," wrote the *Telegraph* on May 29, 1875, the day after the British naval officer George Nares (1831–1915) sailed out of Portsmouth, England, with the ships *Alert* and *Discovery*.[24] The goal was again to reach the North Pole, which remained unconquered, following Julius Payer's attempt. That quote perfectly illustrates how the British saw themselves back then, and the media's ability to inflate expectations. Only the British, and no one else, could conquer the North Pole. Nares expressed some of the same blind optimism that characterized the British expeditions of the early nineteenth century before he set off. He believed that the dangers they would face on their way to the North Pole would be like "child's play when compared with what previous explorers had undergone."[25]

Both ships had three masts, a two-cylinder steam engine, and a crew of 120 men—around the same number of men as Franklin's expedition—sailors, carpenters, smiths, cooks. Most of the officers had no experience of the Arctic, but Nares had been north once before. In 1852, he had gone to look for Franklin, but failed to find him and returned home in 1854. He was supposed to go and fight in the Crimean War, but by the time the ship got there, the war was conveniently over. He sailed through the Suez Canal when it opened in 1869, and on to the Antarctic polar circle, and carried out significant research in oceanography along the way.

Despite the officers' lack of experience in northern climes, *Alert* and *Discovery* managed to sail past Smith Sound and through what is now known as Nares Strait. They sailed farther north than any ship had managed on the American Route. British and American polar explorers had once believed that an ice-free Arctic Ocean was to be found if only the ships got far enough, but the British Arctic Expedition was met by hummocks so tall that Nares was finally convinced that the ice-free ocean did not exist.

The expedition appears to have been as poorly prepared as Franklin's. There was not enough food on board, the men were dressed in flannel and wool, and none of the jackets had hoods. The sleds were too heavy. Most of the crew got scurvy, but not the officers as they had their own healthy food with them in their cabins. No one had skis, but one person had brought snowshoes, and was ridiculed for it. The men, rather than dogs, were going to pull the sleds north. It seems only natural to ask why they did not use dogs, and the answer may be that they felt anything other than man's muscle was somehow cheating. Dogs were too professional and lacked romance—the men had to struggle to reach their goal. Even snowshoes, which would have prevented the men from falling through thin ice, were disqualified as cheating.

The expedition only managed to pass Parry's farthest-north record from 1827 by a couple of miles, but did reach 83° 20′ 26″ N. The plan was to spend two winters in the Arctic, but Nares returned home after only one winter.

Several members of the expedition were already dying from scurvy, and if they had stayed another year, most would probably have died. An extra year away fired people's imagination and expectation, but to come home a year earlier, without having reached the North Pole, was a scandal.

Nares became the first victim of the media in polar history. Journalists who had lauded the men when they set sail now turned on them. Before an expedition, a person was elevated to the status of national hero, only to be subjected to a vicious character assassination when he did not live up to the wildly inflated expectations of the media. It was often the very same journalists who wrote both pieces, and during the buildup and fall newspaper sales rocketed.

Yet the expedition was one of the few that succeeded in carrying out serious research in the Arctic and returned with important and extensive scientific material. Nares was knighted in 1876, received medals, and was eventually promoted to admiral, but very few of the British public cared much about that by then.

The defeat in public opinion was so complete that it became the last expedition to the North Pole to be supported by the Admiralty. Britain's time as a superpower in the Arctic and elsewhere was waning, and in the decades preceding the Second World War the US became the world's leader.[26]

The homage in the *Daily Telegraph* was the last to a polar explorer going north until the British polar explorer Wally Herbert led an expedition from 1968 to 1969, and together with his fellow explorers, Allan Gill, Roy Koerner, and Kenneth Hedges, without a doubt became the first person to reach the North Pole with dogsleds.

5

THEORY AND REALITY

Børge on his North Pole expedition.

A First Attempt to Cooperate in the Arctic

Following a number of Arctic expeditions with scant scientific results, Carl Weyprecht, the captain of *Admiral Tegetthoff*—the Imperial and Royal Austro-Hungarian North Pole Expedition—took the initiative to establish the first International Polar Year in 1875.

A number of nations were interested in carrying out research and learning more about the Arctic. Denmark, the Austro-Hungarian Empire, Finland, the Netherlands, Russia, Sweden, and the US agreed that it was time to work together in the Arctic. Instead of competing to be the first to reach the North Pole, multilateral cooperation would be of more benefit to humanity. Observations over time would be recorded and analyzed in disciplines such as glaciology, cartography, oceanography, marine biology, meteorology, gravitation, and magnetism.

The participating nations established a total of twelve stations throughout the Arctic region in the course of 1881 and 1882, and 700 men were involved in the work. Each nation was responsible for one or two stations and offered participants, equipment, and money. The US chose the most northerly station. The expedition, which was named the Lady Franklin Bay Expedition, was largely made up of American soldiers from the Signal Corps. The Signal Corps had previously been tasked with installing telegraph cables across to the west coast of the USA, which no doubt involved a lot of bad weather, but this would have been very different to what they would face in the far north.

With the exception of two Inuits and one American civilian, none of the twenty-five members of the expedition had any experience of the Arctic.

There was an unusually warm summer in 1881 and there was less ice than normal, so the ship transporting the expedition was able to sail farther north than she would otherwise have done. This proved to be a blessing in disguise. When the expedition reached Lady Franklin Bay on the northeast of Ellesmere Island, they built a house where the men were all going to live for the next two years. The house had three rooms and a floor area of 968 square feet. The camp was given the name Fort Conger, after Senator Omar D. Conger, who had supported the expedition. Even these days, as the world gets steadily warmer, it could be hard to reach Fort Conger via the sea.

The leader of the expedition was Lieutenant Adolphus Washington Greely. Like most of his colleagues from the Signal Corps, he had taken part in the American Civil War. But he had never been on an expedition. Robert Peary, who rarely had a good word to say about other polar explorers, claimed that Greely had told his men that he knew "nothing about ice navigation, I am disadvantaged by having a poor eyesight."[1]

Although the participating countries had all claimed their desire to go to the Arctic was for scientific research, the Americans' only real concern was to get to the North Pole first.

On March 19, 1882, before any scientific trips were ready, and as soon as the sun had risen after the first winter, Greely sent two groups north, one along the east coast of Ellesmere Island and the other along the west coast of Greenland. Both parties had the same goal, to reach the North Pole, or at least to get as far north as they could. On May 13, one group reached 83° 23′; thus the Americans took the record from the British. The new record was around three miles farther north than that set by the almost forgotten English naval officer George Nares in 1876. With imprecise clocks, and an uneven horizon and sextant, I believe the record lies within a reasonable margin of error.

The new record holders returned to Fort Conger on July 1. The Americans had planned that a steamship would arrive the same month, carrying letters, eight tons of supplies, and news from the outside world. All scientific work in Fort Conger was completed, and those who were due to return to the US got ready to travel, while those who were going to winter there wrote letters that could be sent back with the ship.

However, ice conditions were once again normal in summer 1882, and the steamship heading to meet the expedition was stopped by the ice 250 miles south of Fort Conger. The supply party built a depot on land, consisting of coal and food, and left the supplies there, before sailing back.

When the ship failed to arrive, the expedition prepared themselves to winter there for a second time. The sun set on October 15 and rose again 137 days later. In summer 1883, another American steamship, USS *Proteus*, was sent to the rescue. *Proteus* had previously been used in the American Civil War. Unfortunately, she was stopped by pack ice 185 miles south of the station. And what's more, the ship was crushed by ice hummocks and sank before the crew had a chance to build a suitable depot and leave the supplies they had for Greely and his men.

Greely's expedition was forced to spend a third winter there, without new supplies and now totally isolated from the rest of the world. The leader was pessimistic about their chances if they stayed in the same place, and decided that the expedition should leave Lady Franklin Bay and walk south. The captain on *Proteus* had orders to leave the supplies as far north as they could, if they were unable to reach Fort Conger, and this was what Greely knew and hoped to find. Several members of the expedition were against the decision, understandably, as they had enough food and coal where they were. Of course, neither those who wanted to go south nor those who wanted to stay knew that *Proteus* and her load were now 1,300 feet underwater.

The men who went with Greely struggled, froze, and starved to death, if they did not die from scurvy first. One man, Private Henry, was sentenced to death for stealing food. Greely wrote the instructions for how the execution would be carried out in his diary. Three men would be given a rifle each and would then shoot Henry. One of the rifles would have a blank cartridge in the chamber—that way it would not be possible to know who had killed him.[2]

What the instructions did not say, however, was that the expedition only had one rifle left that worked. It is unlikely that Greely had forgotten that they lacked the weapons. He presumably wrote the instructions so the execution would seem orderly after the event and not in contravention of American law. Whatever the case, Private Henry was shot, probably in an ambush. He discovered with time enough to fight back before one of the selected men managed to fire a fatal shot.

THE MEN ATE whatever they had with them or could find to survive—lichen and larvae, as well as boots, clothes, and anything made of sealskin. They also managed to shoot the odd polar bear.

Fresh polar bear meat often tastes like cod liver oil. Børge and I appreciated the taste. Our hunger for fat was such that I could have drunk a bottle of cod liver oil in one go, and I should imagine that Greely's men experienced something similar. We would share whatever fat was left in the frying pan and drink it from our cups, our bodies were so depleted.

The expedition would perhaps have been forgotten, were it not for the six members who survived. In fact, seven were finally rescued, but one died soon after as a result of botched amputations. When they had reached safety, Greely briefly summarized the expedition as follows: "Here we are dying like men. Did what I came to do—beat the record."[3]

When the news that six men were still alive was published on July 18, 1884, it was a sensation. The expedition took up the entire front page of the *New York Times* that day.

On their return to the US, the six men were celebrated like national heroes. They were all offered a promotion in the army. They had fulfilled three essential expectations: They had taken the farthest-north record, beaten the British, and survived to tell of their incredible suffering, which was exactly the kind of news the public wanted.

Once the euphoria had settled, a natural question arose: What had they eaten, in the end, to survive? The *New York Times* had the answer: In addition to the food they had already told of, they were kept alive on human flesh. It was evident that meat had been cut off or scraped from several of the bodies from the expedition, "many of them picked clean" so that only the bare bones remained. The newspaper claimed one or more of the men had been killed for food. The news was met with disgust, even by those who felt it was acceptable to eat a dead friend when the alternative was to die of starvation. In a leading article, the newspaper concluded that the men were driven by "horrible necessity to become cannibals" and the expedition was the most "dreadful and repulsive in the long annals of Arctic exploration."[4] The American authorities knew of the circumstances, but, according to the paper, had decided to withhold the information. Greely was adamant that he knew nothing about the cannibalism and the newspaper's claim that people had been killed for food was never proved.

It is hard to know in retrospect what each of the six men ate. Whatever the case, it would be hard to hide if you were taking meat from dead bodies and eating it when you live in close quarters with others who are on the point of starvation. Even if you managed to hide what you were eating, it would only be natural for others to wonder why you were not getting thinner and thinner like them.

For me, cannibalism is a kind of necessary evil in situations like that, and therefore acceptable. In the 1980s, early on in my career, I gave considerable thought to what I might do when the standards of the civilized world are no longer relevant. I would certainly eat a friend who was already dead if the alternative was to starve to death. It would of course be terrible to be forced to do this, but when the alternative was even more brutal, I think my companions would understand.

The *Jeannette* Expedition

On July 8, 1879, the American three-masted steam bark *Jeannette* sailed north from San Francisco. She was to sail through the Bering Strait and then drift and sail to the North Pole. Like most other ships that had sailed to the North Pole, she was not built for Arctic conditions, so her hull had been reinforced before leaving.

Like many previous expeditions, this route was based on the theory of an ice-free polar sea.

The *Jeannette* expedition was sponsored by the newspaper baron James Gordon Bennett Jr.; he had contacted German cartographer Petermann in 1877 to seek his advice. Petermann explained to Bennett that "if one door will not open, try another."[5] And so he had. Petermann still believed that there was a warm climate and diverse animal life at the Pole. But he also believed that he had found a new way there—the Kuroshio Current, or Japan Current, which runs along the east coast of Japan, through the Bering Strait, and on into the Arctic Ocean, where it meets the Gulf Stream at the polar point. Petermann believed that it was thanks to these ocean currents, which are both relatively warm, that the Arctic Ocean was ice-free.

This was an era when America's most powerful newspaper barons organized expeditions to have exclusive access to the stories. The recipe for Bennett's newspaper, the *New York Herald*, was innovative and simple: All news should be sensational. And he was right. Murder, suicide, fire, and men in danger of dying on the ice sold more newspapers than good news. Bennett, who was not only the owner of the newspaper, but also the editor in chief, had already organized rescue operations to search for human remains and equipment from the Franklin expedition. To ensure exclusivity, a journalist should not always wait for things to happen, but rather create the news.

The newspaper baron was also the funder behind the Welsh-American Henry Morton Stanley's expedition to rescue the Scottish missionary David Livingstone in Africa. Livingstone was obsessed with finding the source of the Nile, a dream that had existed since Herodotus wrote his *Histories*, and the missionary had not had contact with anyone outside Africa for six years. No one in the US or Great Britain knew if Livingstone was still alive. And even if he was dead, there was no reason to abandon the expedition, according to Bennett. His instructions to Stanley were clear: "Bring back all possible proofs of his being dead."[6]

Livingstone never found the source of the Nile, but he was alive when Stanley reached the village of Ujiji by Lake Tanganyika on November 3, 1871. The missionary and explorer, ill and toothless, was sitting on a veranda with Arabic merchants when Stanley rode up on a donkey. He had no wish to be rescued. (Stanley later omitted anything about that part of the story.)

The article telling how Livingstone had been found traveled for eight months, from Zanzibar to Bombay, where it was telegraphed to London, and then on to New York, before it was published in the *New York Herald* on July 2, 1872. The article opened with: "Dr. Livingstone, I presume." The pages about their meeting have disappeared from Stanley's diary, apparently torn out by Stanley himself. It is doubtful that the words were ever spoken aloud.

Bennett's goal was to engineer more moments that sold as many newspapers as the meeting between Livingstone and Stanley. Of all possible destinations for an expedition, he believed that an expedition to the North Pole was the surest way to create an equally splashy news story. And he would be proved right.

The US naval officer George W. De Long was engaged to lead the expedition. He had a wealth of experience of voyages to and from Europe, the Caribbean, South America, and along the east coast of the US. The *New York Herald* had already made him a hero in the US in connection with one of the newspaper's expeditions to rescue Emil Bessels and other members of the *Polaris* expedition that had not been found on the ice in summer 1873. Early on as he traveled north, he wrote to his wife, Emma: "I cannot help thinking how much happier we should be if we were together." He promised that when he came home, he would take a year off so that he, his wife, and their daughter could go on a long holiday.[7]

De Long did not find any members of the *Polaris* expedition—he was unaware that the last members of the team had in fact been rescued as he sailed north. However, the expedition was not entirely wasted for the naval officer. A journalist accompanied the expedition along the coast of Greenland, and wrote about De Long's courage and leadership, and how he saved the lives of everyone on board on his rescue expedition.

According to Emma, when De Long returned home to his family late that summer, he was a changed man. The idea of a long holiday was set aside. Emma saw something different in his eyes; it was as though he had a fever, as though all he wanted to do was go back. "The polar virus was in his blood and would not let him rest."[8]

IN SEPTEMBER 1879, the *Jeannette* expedition reached the pack ice and, to the crew's surprise, the ship got frozen in the ice. They could not sail further. The ship remained locked helplessly and they drifted slowly northwest. De Long wrote in his diary that while it might be thrilling for those back home to read about wintering in the Arctic sitting at home in front of a warm fire, the reality was very different, "sufficient to make any man prematurely old." He described the sound of ice hummocking very well: "a rumble, a shriek, a groan and a crash of a falling house."[9] After twenty-one months in the ice, they still had not seen any sign of the Kuroshio Current that Petermann had said made the Arctic Ocean ice-free. Which is not so strange, as the Kuroshio Current swings east when it has passed Japan, before reaching the Bering Strait and Arctic Ocean. On the night of June 11, 1881, the ice pressed so hard against the hull of the ship that it splintered. In the moonlight, the men on board watched two enormous ice floes crash into each other, and the ice around the ship started to move. Their position was 77° 15′ N, more than 870 miles from the North Pole.

All members of the expedition survived the shipwreck, but several of the crew almost drowned. When the hull of the *Jeannette* started to be crushed by the ice, the officers realized what was about to happen and packed their bags. The noise of the ice floes hummocking with such force must have been chilling. Apparently, De Long asked the officers to be discreet, and not let the crew see that they had packed their belongings.[10] Even if it put their lives at risk, he was more concerned that the crew should continue getting the supplies and equipment off the boat.

The expedition might possibly be the best financed and least successful of all North Pole expeditions.

THE THIRTY-THREE MEN who had been on board *Jeannette* started to walk south over the ice toward the Lena Delta in northern Siberia. They pulled three smaller boats with them. The plan was that when they reached open water they would row or sail on, with fourteen men in De Long's boat, eleven in the second boat, and eight in the third. The men were driven mad by hunger and thirst. One had syphilis in his eye and several suffered from gangrene, as a result of frostbite, poor blood circulation, and dead tissue. Gangrene causes the flesh to rot on the body: extremities such as hands, feet, and nose are most at risk. If amputation is required, it is often incredibly urgent as the disease can spread, and should be done farther up the infected limb to ensure that all the gangrene is removed. So, amputations were carried out on the move and bones were left exposed.

I still remember reading about polar explorers and gangrene when I was a boy. I knew what the rotting flesh of a cow or pig smelled like, but was surprised when I learned that rotting human flesh smells the same. The smell was upsetting not only for the person who was suffering, but also for anyone sharing their tent. I no longer remember who it was, but a polar explorer I read about when I was young, told how he had himself found a pointed stone, put his foot flat on a flat slab, and chopped off his own toes. He said that it was psychologically difficult to chop off bits of yourself, but it was a relief to be rid of the smell.

Twelve of the fourteen men in De Long's boat died. All eleven in the second boat survived, and the third boat got caught in a storm and none of the men were ever seen again. Bennett saw great potential for more headlines as time passed and the expedition did not return. The possibility of repeating the success of the Stanley and Livingstone story inspired him to send several small rescue operations north, with journalists. These operations made for sensational copy.

A journalist from the *Herald* finally found De Long and eight of his men. They were all dead. Those who had died first had been buried by the others, before they then also perished. The journalist dug up the graves. The bodies were examined for signs of violence, cannibalism, and murder, but no evidence of the sort was found. De Long's widow, Emma, described the journalist's behavior as "the bitterest potion I had to swallow in my whole life."[11] The bodies were sketched. A number of letters and diaries were found, written by De Long and other members of the expedition.[12]

De Long's last entry in his diary read: "October 30th, Sunday— One hundred and fortieth day. Boyd and Görtz died during night. Mr. Collins dying."[13]

SOME OF THE men sent by Bennett to rescue survivors from *Jeannette* died in the struggle, which again gave the newspaper great headlines. Survivors from the expedition also went back to save their colleagues, providing the newspaper with more good material. The *Jeannette* expedition created even more articles than Stanley and Livingstone and was a huge success for the newspaper.

There are many reasons why the expedition ended in tragedy. The boat was not suitable, the crew lacked experience, and they sailed from San Francisco too late in the summer. At the last minute, they were also told to look for the Finnish-Swedish explorer Adolf Erik Nordenskiöld (1832–1901), leader of the *Vega* expedition. The *Vega* expedition was on its way through the Northeast Passage and did not need help, but Bennett envisaged a kind of Stanley–Livingstone moment. In the end, De Long chose not to take the major detour necessary to find Nordenskiöld, but it must have been frustrating all the same to suddenly be given new instructions when your goal is to be first to the North Pole.

The US Navy ordered an inquiry as the media and families demanded to know why the expedition never got anywhere near the North Pole—and why so many died. The conclusion was that De Long was not at fault. The Collins family, who were mentioned in De Long's last diary entry, protested. Congress then investigated the expedition and drew the same conclusion as the navy. I can only presume that this was a given, as the media, navy, and Congress all wanted the expedition to be seen as a success.

The most important scientific result of the expedition was that the several-thousand-year-old theory of an ice-free Arctic Ocean and temperate zone around the North Pole was finally put to bed. August Petermann, who never completely abandoned the theory, did not hear the outcome of the expedition as he took his own life while the rescue operation was ongoing.

ONE REASON WHY many men choose to join expeditions, which is not often discussed, is the desire to get away from family life. To escape the children and marital bed. As Fridtjof Nansen knew, it is often easier to appreciate your family when they are at a great distance. That does not appear to have been the case for De Long. To start the expedition from San Fransisco, De Long first had to sail the *Jeannette* from Le Havre in France across the Atlantic Ocean, round Cape Horn, and north up the Pacific Ocean to San Fransisco. The voyage lasted eighteen months and he took Emma with him. "Well, your wife must think a great deal of you" was Bennett's comment. He added, with perhaps a little envy: "No woman would ever do that to me."[14] When De Long then set sail north from San Francisco, Emma wrote beautiful letters to him, which were sent on boats heading north, in the hope that one or more might reach her husband. The first letters were full of optimism, joy, and yearning, but they gradually became more resigned. All are full of love. Of course, her husband never received any of the letters. He, for his part, wrote of his wife with love and warmth in his diary, which she was able to read after De Long had been found dead.

American Freak Shows on Ice

From 1898 to 1905, the media was a driving force behind several almost forgotten American expeditions that started from Franz Josef Land, the northeastern coast of the North Pole closer to Russia, rather than following the American route. The only book written about these expeditions is called *The Greatest Show in the Arctic*, by the anthropologist P. J. Capelotti. For the reading public, the expeditions were a kind of freak show—where Franz Josef Land became an arena where people went missing, starved to death, and died without any hope of getting to the North Pole. Very little of any scientific value came out of them. But the stories were loved by readers and lecture audiences. Politicians and the media could bask in their glory or complain about them to draw attention. Newspaper and book sales increased accordingly.

The American industrialist William Ziegler (1843–1905) backed three of the expeditions. Ziegler was not tight with the purse strings, and the first expedition is said to have been the best equipped since Franklin's final attempt in 1845, with 428 dogs, forty-two men, and fifteen ponies. Ziegler made his money from baking powder, and like so many others who have found success in one field, it gave him immense confidence in other areas as well. He was convinced that he knew how to get to the North Pole. But he had no desire to go there himself, so got others to do so on his behalf and to name after him any new countries and islands they might discover. (Ziegler exemplified the saying "A rich stuntman hires his own stuntmen."[15])

Once the expeditions had anchored at Franz Josef Land, a number of attempts were made to reach the North Pole. The distance to the North Pole was 560 miles, and on one attempt they only managed to cover .3 miles. The reasons for the lack of progress were always the same. As someone on board one of the ships said: "The polar party was back in a week with the same old tale of impossible obstacles—storms, open water, failure."[16] Even nowadays, the most frequent explanation given by adventurers who do not reach their goal is the weather and open water.

WALTER WELLMAN, AN American who left his career in journalism, criticizing and ridiculing those in power, in 1892, tried to find Columbus's original landing place in America. His new goal was to create his own headlines as a polar explorer.

Wellman organized five expeditions in his bid to "make a dash" to the North Pole, as he put it. In three of his attempts, he was to use his own legs, and for the final two he intended to fly an airship. A newspaper baron in Chicago financed two of these expeditions on the condition of exclusive coverage.[17] Yet none of the expeditions got much farther than the boats that had transported them to the edge of the pack ice. His second expedition, in 1888–89 was his most successful in terms of how far north he got.

Wellman's hope was that the ice would be flat until they reached open water closer to the Pole, and then that the weather would be good enough to row and sail to the final goal. He claimed that Fridtjof Nansen supported his plans, but the truth was that Nansen felt nothing but contempt for "what he saw as American materialism, commercialism and lack of culture."[18]

Like so many polar expeditions, Wellman and his men loaded their three sleds with so much that they could only move them one at a time, and so had to go back and forth. This meant that for every mile they progressed north, they had in fact walked three. It was wildly inefficient. Wellman had some Norwegians with him, who unlike many of the others on board were paid wages. Wellman was critical of Peary and believed he had made a mistake relying only on Inuits, as according to him, Norwegians had more technical knowledge, greater resilience, and were "infinitely more intelligent and loyal."[19]

The Norwegians were accustomed to an expedition culture whereby everyone was equal and mucked in together, whereas the American leaders gave the paid Norwegians all the hard work. Two men, Paul Bjørvig and Bernt Bentsen, were ordered to guard the expedition's northerly depot, while the rest of the expedition lived more comfortably forty-one miles farther south in the archipelago. Bentsen had crossed the Arctic Ocean in *Fram* with Fridtjof Nansen and was known to be a good storyteller, which no doubt helped over those long three years. Bjørvig and Bentsen built a hut, or a cave as they called it, that was sixteen feet long, with three-foot-high stone walls. The chimney, which went out through the roof, was made from tin cans. Their supplies for the winter were ten frozen walrus, some canned food, blubber to heat the primitive hut, four candles, and one newspaper. They had no wicks, so made them from dry moss.

During a polar night, the days and nights become alike. The two settled into fixed routines with two meals a day, feeding the dogs once a day; they read and reread the newspaper, and soon got fed up reading what was written on the tin cans. They spent time maintaining the hut, and lay looking at the ceiling, which was made from walrus hide. On some occasions polar bears started to eat off the walrus hide and Bjørvig fired a shot with his rifle through the hide to scare them away.

The cold took its toll. As early as November, Bentsen fell ill and started to lose sensitivity in his body. Gradually the blood stopped circulating, first in the feet, later in the calves and thighs, and finally farther up. Closer to Christmas Eve, Bjørvig wrote that his body was as white as snow, and he added that Bentsen never complained.

Between Christmas and New Year 1898, Bentsen started to sing the Norwegian Christmas song "The Earth Is So Beautiful." Perhaps he thought the familiar psalm would cheer them up, but it was unnerving for Bjørvig to hear the weak voice of his friend inside the hut. It strikes me, when I read Bjørvig's diary, that Bentsen was singing his own funeral song as he lay there, spent, in the dark hut, with the polar night outside:

To Paradise we walk in song
Ages shall come
Ages shall pass

Bentsen became delirious before he managed to sing the last verse of the song. On the morning of January 2, Bjørvig woke up and felt that Bentsen was dead. It was as always dark inside; he lit a match and saw that his friend was no longer alive. In the autumn the year before they had agreed that if one died, he should not be left outside the hut to be eaten by polar bears and foxes.

Bjørvig melted ice and washed Bentsen's hands and face and left him lying in their shared sleeping bag. The Norwegian wrote that it was sad to lie so close to a dead man. It was cold enough when he was alive "and it feels worse now he lies there dead. I have to take it as it comes." Bjørvig often thought that Bentsen was the lucky one, because he did not have to freeze anymore.

After Bjørvig had spent fifty-five nights in the same sleeping bag as his dead friend, Wellman appeared with two other Norwegians from the expedition. They were on their way to the North Pole. They buried Bentsen in the bare, frozen ground, covering his body with stones. Bjørvig made a cross with two bits of wood, inscribed Bentsen's name, and mentions in his diary that it must be the northernmost grave in the world. The day after the funeral he went north with the others. Three days later, Wellman injured his leg and they had to turn back. Bjørvig wrote that "the whole journey was wasted."[20] In his book, Capelotti adds that the three Norwegians were intent on keeping Wellman alive until they reached the rest of the expedition. It would not look good if the only American did not return.

Leo Tolstoy's Polar Explorer

When I think about all the suffering that North Pole explorers have endured and the way in which Bjørvig describes his days and nights with Bentsen, his stoicism reminds me of the experiences of Pierre in Leo Tolstoy's *War and Peace.* For the greater part of the book, the aristocrat Pierre lives an unexciting and monotonous life of luxury, but then he is taken prisoner by Napoleon's army, and as a prisoner of war is forced on a march in which two-thirds of the men die. But rather than being a catastrophe, the march instead rescues him from a life of futility.

Parts of Pierre's life start to resemble that of a polar explorer; his boots are worn out, his feet ache and are covered in open, infected wounds and scabs, the men are given only a little horse meat to eat. Those who are too weak to walk die by the roadside, and Pierre experiences a stoic calm he has never felt before. You suffer more in the mind than in reality; it is the perception of life that is decisive. It is not so much what happens, but how you respond to it.

Tolstoy's Pierre learned not only with his intellect but with his entire being that happiness comes from within, and dissatisfaction can too. When he had put on tight dancing shoes he had suffered just as he did now when he walked with bare feet that were covered with sores—his footgear having long since fallen to pieces.[21]

Tolstoy did not write about polar expeditions, but polar explorers might recognize themselves in his descriptions—the pleasures of simplicity, the food that at first tasted disgusting that Pierre then comes to appreciate. That suffering, great and small, appears to be constant, whether one is at home in a warm room or out on the ice. Børge and I often talked about how short the distance was between comfort and security, and danger and desperation.

6

THE HEROIC ERA

Polar explorer Fridtjof Nansen and his wife, Eva.

Fridtjof Nansen—A Well-Educated Stone Age Man

It was only when I was preparing to go to the North Pole, two decades after I was given my first globe, that I understood it was the cold and the danger that made it meaningful. If it had been simple and risk-free, I would never have dreamed of going there.

The most northerly part of the world has never attracted people who recognize the difficulties they will face there; it has been the domain of the naive. Astronomers, writers, geographers, meteorologists, marine scientists, zoologists, sailors, artists, crooks, rogues, adventurers, and fortune hunters—most polar explorers fall into two or more of these categories. Only a few found fame and fortune. Most of those who survived returned home to obscurity and poverty—some carrying crushing secrets of trickery, cannibalism, and murder. The fates of these explorers were played out while their wives and children waited at home, unaware, sometimes for years.

ONE WRITER, A zoologist and artist who returned home safely and found fame and fortune was Fridtjof Nansen. Thanks to his expeditions across Greenland and to the North Pole, Nansen became one of the most famous people in the world. In contrast to many others, he did not make his name in battle; nor was he born into fame, like princes, counts, and barons.

A new technology developed while he was in the wilderness, which made it easy to distribute high-quality photographs across the greater part of the world within seconds. Newspapers and books were suddenly full of photographs, and this benefited Nansen more than others. Few heroes have been as photogenic as he was. I think he spent more time than other polar explorers in front of a mirror, studying how to pose for the camera. His clothes had to fit perfectly. When he was not happy with the cut and fabric of his clothes, he made his own outfits.

He managed to highlight the features that were most important to him—his height and muscles, his defined cheekbones—and he often wore tight-fitting trousers and jackets to emphasize his broad shoulders, elegant posture, and firm buttocks. His good friend Erik Werenskiold, the artist, chose Nansen as the model for Olaf Tryggvason, the Viking king, when he illustrated Snorri's Heimskringla, the Norse king sagas, toward the end of the nineteenth century.

Nansen also used a self-timer to take pictures of himself without clothes on. In one photograph, he is at home, lying on his back in his white bed. In another, he is standing up, flexing his muscles and straight back. And in the third, he is lying on his side looking into the camera. The private parts of his body are not covered. He sent the photographs to a woman he was in love with in the US.[1]

Unlike other polar explorers, Nansen used his fame to campaign for causes he believed in, such as Norway's independence from Sweden and international efforts to alleviate the famine in Russia.

Nansen's thoughts on how to reach the North Pole were simple, and close to my own. You have to fight *with* nature. If you fight against nature, you will never get there, or as he put it: "You should not work against the current, you should go with it."[2]

For almost 400 years, from the time that the Englishman Robert Thorne set sail for the North Pole to when Nansen sailed north in 1893, expedition leaders all made the same mistake: They believed there was an ice-free ocean. They not only fought against the current, but also the cold, the ice, and sometimes even common sense. Nansen did not believe that the Arctic Ocean was ice-free and so he acted accordingly: He sailed into the ice and allowed the boat to get stuck.

MUCH OF THE history of the North Pole is about men who were not willing to learn from their predecessors' mistakes, but Nansen understood how ridiculous that was. He loved to quote one of his favorite philosophers, Thomas Carlyle, a Scotsman: "Experience is the best of schoolmasters; only the school-fees are heavy."[3] Nansen studied De Long's tragic *Jeannette* expedition. A pair of waterproof trousers bearing the name of one of the sailors, Louis Noros, drifted 2,900 nautical miles from where the ship had been crushed by ice on the other side of the Arctic Ocean to Greenland. Nansen asked himself which route the drifting equipment had taken? And his answer was: the shortest way, over the North Pole, "on an ice floe and the ice floe drifts on the current." Noros fortunately managed without his waterproof trousers: He was one of thirteen who survived the *Jeannette* wreck.

Nansen chose to break with another tradition. Instead of adapting an existing ship, his ship would be specially designed and built to deal with the ice. The motorized schooner, *Fram*, was an ideal ice-ship, small and strong and shaped in such a way that the ice could not crush her, as it had done with *Jeannette* and *Polaris*. "I will build a vessel with a rounded hull, so that when the ice presses into it, it will not get a hold and crush it, but will instead slide . . . and lift the vessel up."[4] He installed a windmill on deck so the expedition could have electric lighting through the winter. It produced enough electricity for 20,399 light hours, and they had light for the first two winters. (The windmill was then dismantled to make sleds and other equipment.)

But Nansen also broke with a third tradition, which is not so widely recognized. One of his primary concerns was that everyone on board should survive. No longer would human life be of so little value in the vicinity of the North Pole. Nansen did a lot of research to ensure that they had nutritious food on the expedition. He contacted leading experts in England, and the nutrition and fat in all the food were measured. Dogs and maps were bought from Russia.

But there were other things that Nansen did not see as important. For example, he gave little thought to the fact that the men might need some enjoyment when they were stuck in the ice. No one got more than two cups of coffee a week, there were no birthday or Christmas presents, and he introduced a smoking ban indoors. This was lifted on Christmas Eve the first winter, when he realized he wanted to lie alone in his cabin and smoke a pipe.

AFTER HE CROSSED Greenland in 1888, Nansen had the opportunity to learn about living in the Arctic from the Inuits. He arrived at Godthaab, or Nuuk as the town is now called, after the last boat of the year had sailed back to Europe. So instead of returning to Norway in the autumn, as planned, the members of the expedition had to stay there through the winter. And because the last boat had sailed, no one was able to write home to say how the expedition had gone.

Nansen did not waste his time there. He learned to speak the language and was soon convinced that the Inuits' way of life and culture were superior to that of Europeans. I would tend to agree. The Inuits that Nansen got to know derived immense pleasure from very little. It frustrated Nansen that the opposite was true in the societies he knew. Not only did people take so little pleasure in so much, in our so-called civilized part of the world, but we also seemed to be turning our back on nature.

Nearly everything was made locally, and the Inuits used more or less 100 percent of anything they hunted or caught. When the Inuits killed a walrus, they ate the meat, used the skin to make kayaks, made oil from the blubber, and created tools from the bones. Even in Nansen's day, people in the West had a much bigger carbon footprint than the Inuits, and in modern times our footprint is countless times bigger than the footprint of the Inuits back then.

Nansen moved in with the Inuits for the winter. He learned how to hunt like them, and after a while could kill a seal, walrus, and polar bear standing or lying on the ice, or sitting in a kayak. He also learned how to use everything from the animal, and how to make a copper oil lamp, effectively a combined cooker and lamp; Nansen called the metal, a copper alloy, "new silver." These skills would save the lives of Hjalmar Johansen and Nansen in the winter of 1895–96.

NANSEN BELIEVED THAT the Arctic Ocean held the answers to many key scientific questions regarding oceanography, gravity, magnetism, the aurora borealis, flora, fauna, meteorology, marine geology, and climate history. He had a good grasp of all these fields. But I admire Nansen for so much more. Not only was Nansen an important scientist and polar explorer, but he was also a great humanist and was awarded the Nobel Peace Prize in 1922 for his humanitarian work. He was also an important diplomat for the Norwegian foreign office. He was a fine artist and could draw a polar bear attack in the Arctic Ocean; a competent photographer with an eye for dramatic compositions; and a good storyteller and author.

If Nansen was going to find the answers to those big questions and share them with the world, no one but him, according to himself, could lead the expedition, even if this meant that he had to leave his wife and daughter for an indefinite stretch of time.

AN INTEGRAL PART of expeditions north is often longing for the wife and children you left in the south—and their longing for you. Eva Nansen wrote that when *Fram* sailed from Kristiania, as Oslo was then known, with supplies on board to last five years, "I lay in bed, . . . and wept and wept."[5] Their daughter and first child, Liv, was only six months old.

I can understand why Eva cried, and equally I understand why Nansen could not give up his dream of the North Pole. The Pole was an obsession, and Eva was not. This is often a dilemma for couples when one of them yearns to go to the North Pole. The other is attracted to the adventurer who nurtures an intense desire to be in nature for long periods of time, who is bursting with curiosity and wants to come face-to-face with the sublime—the very characteristics that will eventually drive them apart. The relationship is difficult when they are together, and as time passes, it becomes difficult when they are apart.

Even though the North Pole appears to have been more important than his marriage, I do believe Nansen when he describes his feelings for Eva, as he finally sets sail for the North Pole. It is easier to love when you are going to live on the ice indefinitely, because that love makes no demands. Nansen wanted what he did not have; it was impossible for him to be content.

He finished his last letter for three years only minutes before the ship set sail along the Northeast Passage: "I carry you with me everywhere, through the mist and ice, over all the oceans, as I work and in my dreams, you are everywhere." He continues: "Let my last word to you be: don't worry about me, whether I am away for a long or a short time . . . *there is no danger*, absolutely none at all."[6] Then he disappears. His daughter Liv later told how the boat's departure was delayed until he had finished the letter.

It was important for Nansen to convey that he was not going to the North Pole simply for his own sake, but rather for research. Back then, as now, science was a more honorable motivation than adventure and unabashed self-promotion. Nansen maintained a long tradition among polar explorers of hiding the more egocentric reasons for journeying to the North Pole. Few have been as refreshingly honest as Robert Peary. He openly stated that fame and fortune were his main motivations.

In my experience, any initiative to walk, sail, drive, or fly to the North Pole has been driven by the need for adventure and recognition. But I fully understand why Nansen highlighted research: He was a scientist, and the expedition needed financial support from the Norwegian parliament, the king, and rich Norwegians (adventure and personal gain were not enough to secure their backing).

Later in life, Nansen was more open about his motivations. In his book *In Northern Mists*, he quotes from *The King's Mirror*, a philosophical and didactic text written for the sons of Haakon Haakonsson, the Viking king. The form is simple: The son asks and the father answers. The quote that Nansen chose is about the three incentives for a man to travel out into the world. As a reader, it seems natural to assume that the words also apply to Nansen and how he understood himself:

> One part of him is emulation and desire of fame, for it is man's nature to go where there is likelihood of great danger, and to make himself famous thereby. Another part is the desire of knowledge, for it is man's nature to wish to know and see those parts of which he has heard, and to find out whether they are as it was told him or not. The third part is the desire of gain, seeing that men seek after riches in every place where they learn that profit is to be had, even though there be great danger in it.[7]

One of Nansen's biographers, Roland Huntford, compared him with Johann Wolfgang von Goethe's story of Faust. Not because Nansen entered a pact with the devil, but because he shares Faust's ambition to discover how the forces of nature work together. They also both felt they would be doomed if they were happy for too long. A keen observation, as Nansen read Goethe's *Faust* and, as Huntford mentions, expressed much the same thing: "Goethe's Faust never reached a place where he wanted to 'remain.' I cannot even glimpse anywhere worth the attempt."[8]

Nansen used his passion and fame to drum up international aid during the great famine in Ukraine, which was then part of Russia, from 1921 to 1923. Twenty million people were starving and unspeakable numbers were dying. He was one of the few Western European leaders who went to great lengths to help, yet for the rest of his life was deeply unhappy that he had not been able to do more.[9] A copy of Goethe's *Faust* was taken on board *Fram*, not as part of the ship's library, which contained 600 books, but in Nansen's own luggage.

No Possibility of Retreat—Three Times

In the nineteenth century, expeditions were all planned with a line of retreat for the ship and crew. The British, Americans, Germans, Austrians, and Hungarians all chose to sail north as close to land as possible.[10] Nansen, however, maintained that a line of retreat was a snare. If you have no choice but to continue, "then you have either to persevere or to perish."[11]

One reason why Nansen succeeded in crossing Greenland in 1888 was that the members of the expedition knew they were likely to die from starvation, cold, or fatigue if they turned back. In 1886, Robert Peary started from the west in his attempt to cross the island. This meant that the participants knew that at any point they could go back to the settlements and Egypt's warming meat stews.

Nansen was true to his philosophy when he headed to the North Pole with *Fram*. As soon as the ship was trapped in the ice and started to drift north, Nansen and his men had no line of retreat. The name that Nansen chose for his North Pole boat is a reflection of this one-dimensional philosophy: "Our hope lay ahead and so the ship was called *Fram*, which means 'forward.'"

In the event that *Fram* was wrecked, Nansen assumed that they would be able to drift on a large ice floe until they reached open water. He cited the catastrophic *Polaris* expedition, which ended up drifting south on an ice floe for 196 days, as an example that it might work. I am uncertain whether he said this to impress people, or if he actually meant it. The aforementioned Thomas Carlyle defined a genius as someone with an inexhaustible ability to never give up. Nansen quoted him on this, and added that given time and patience, mulberry leaves can be turned into silk.[12]

The Long Journey South

On board *Fram*, the hope was that they would drift with the pack ice right over the North Pole, but after they had been frozen into the ice for eighteen months, it was clear to Nansen that they would drift a couple hundred miles south of the Pole. This was not close enough, and as he explains: "My sense of adventure realised more could be done."[13] Nansen was the only scientist on board, but on March 14, 1895, he left *Fram*. He wanted to be the first person to reach the North Pole—the pursuit of knowledge was now openly of less importance.

Two people left the safe embrace of *Fram* that day in March. The other person was the Norwegian gymnastics champion Hjalmar Johansen (1867–1913), who had joined the expedition as the stoker, and according to his contract was willing to follow all "orders and take on all manner of work."[14] Johansen was fit and had impressed everyone on board by turning a somersault in full polar gear, without preparation. I have never heard of anyone else who managed that. They had three sleds with them that each weighed 485 pounds, in addition to their kayaks, which were balanced on top of two of the sleds. The sleds were pulled by twenty-eight dogs, and after some time, any exhausted dogs became food for the others. The dogs initially showed "great disgust at eating their comrades," Nansen writes, but soon found it easy enough as their hunger increased. Humans and dogs become more and more alike as time passes on the ice. They had enough supplies for three months, but the journey took fifteen months in the end.

It is strange to think that no expeditions before Nansen had thought of using skis. Those of us who can ski all know something the other polar explorers clearly did not: that it is far easier and faster to slide over the snow than to walk on foot. Nansen had been skiing since he was two years old. The sleds did not weigh as much as on previous expeditions and the wooden runners caused less friction. Johansen and Nansen also had a third, major advantage: the kayaks. These were considerably lighter than the boats that had been dragged over the ice on earlier expeditions, in the hope that they would find an ice-free ocean. Minimizing weight is crucial in reaching the North Pole. Dogs, sleds, kayaks, and sealskin clothing—Nansen had learned a lot from the Inuits.

AS THE TWO men progressed north, they struggled with the same challenges as expeditions before them: the cold, leads, and hummocks. On April 7, they set the record for reaching farthest north at 86° 14′ N—260 miles from the Pole itself. Nansen was pleased with the record, and decided they should turn around and head south. Although he doesn't mention it in his diary, he must have been aware that even their record would be eclipsed, and only the first person to reach the North Pole was likely to enjoy lasting glory.

We too battled against the same things. The day we set up camp at 86° 13′ N was Børge's son's birthday. To celebrate, Børge had smuggled a bag of dried fruit soup with him, which needed to be heated with a quart of water. We used three quarts and shared it equally. Børge took out two photographs of Max, the only personal belongings he had with him. My personal belongings were a tube of toothpaste and the New Testament (the idea being to take a book that contained as many thoughts as possible per gram: It weighed .98 oz., but I never had the energy to read it). Børge wrote in his diary: "Sadly, I am drifting around in the Arctic Ocean right now instead of playing with you and teaching you the things that a proper father would." He promised himself he would do that as soon as we got home. Even before we left, he had been criticized for choosing the North Pole over his family. There was no time in the tent for such thoughts. We agreed the criticism was unfair, and that it would be better for Max, all round, to have a father who chose his own way and followed his dreams. I wrote: "It's a great way to raise children, in the long run." Today, I know that we were right, providing the expedition ends well.

We talked a little about Johansen and Nansen and their record, but were too tired and hungry to think about anything other than practicalities and food. We were both suffering from malnutrition, it felt like the fat was melting off our bodies, and our hungry stomachs had started to wake us at night. "Hunger is shadowing us like a ghost," Børge said. We talked about pasta carbonara, blueberry pancakes with syrup, and Børge's favorite cake. A few days later, Børge wrote in his diary that these were "the toughest days I have experienced since the day I was born."

IT IS NANSEN and Johansen's torturous return journey south from 86° 14′ N that fascinates me most. A century before GPS became commonplace, it was almost impossible to find *Fram*, as she had drifted on over the Arctic Ocean. They therefore planned to return to Norway via the Pole under their own steam. For the third time, Nansen did not have a well-planned line of retreat.

Most people would consider this irresponsible, deliberately increasing the risk of death. But for Nansen it was different: The journey became one of his greatest triumphs. One difference between a genius and the rest of us is that a plan might work in one situation precisely because the genius constantly comes up with new solutions. Nansen possessed an optimism that others might think naive. He romanticized being left behind to live on an ocean which "already in April is teeming with bears, auks and guillemots, and where the seal lies on the ice, this must be a Canaan, overflowing with milk and honey."[15] The same ocean that for centuries had claimed countless lives.

Nansen's plan was to walk to the west side of Petermann Island, which was discovered by Payer and drawn into the map in 1874, and then kayak to Svalbard. Payer had only seen the island from a distance, and Nansen's map showed a sketch of what Payer had assumed was the south end of the island. The north end was not included because no one had ever seen it. Petermann Island had been positioned roughly one latitude north of Franz Josef Land, which was drawn in more detail. Unfortunately for Johansen and Nansen, what Payer had seen was in fact a mirage, and the island, as with another of Payer's discoveries, King Oscar Land, did not exist.

In his diary, Nansen speculated whether Payer might have lied about Petermann Island, in which case he was glad the lie had "cost us no more than a year's delay." It seems unlikely that he lied. It is often hard to differentiate between an iceberg, a cloud, and something that could be an island in the middle of the Arctic Ocean.

Payer later claimed that the reason why the two had not found Petermann Island was that they were lost, as their watches had stopped and they were not able to establish where they were. Payer was right about their watches: Nansen and Johansen extended their days on their way south and at times were awake for thirty hours straight. On one occasion, Nansen, who was in charge, and Johansen forgot to wind up their watches.

Børge and I also changed our daily rhythm as we made our way to the North Pole, and toward the end could be awake for twenty-seven hours straight and then sleep for eight. The reason for this was that Ran Fiennes, the English explorer, was leading an expedition from Russia to the North Pole, and we wanted to *beat the English*. Ran had already completed a round-the-world expedition via the North and South Poles, and had the experience that we lacked. I was so exhausted that at times I felt like I was fainting, my fingertips were frozen for days and completely white, the blood circulation long since lost. Common sense told us we should sleep more, but as we kept reminding ourselves, you do not get to the North Pole by making only sensible decisions. A polar explorer exists at the very limits of reason. Accordingly, we should act impulsively, and at times ridiculously, to embrace the opportunities that life offers.

After my own experience of extending the days, I find it incredible that Nansen and Johansen remembered for so long, despite lack of sleep, to wind their watches at the right time for as long as they did.

When Nansen and Johansen no longer knew the time, they had to navigate by the sun, moon, compass, and dead reckoning (when you calculate yourself how many miles you have covered in which direction and at what speed). When the ice is undulating, you might manage around one mile per hour, so that if you have been walking for twelve hours, you have progressed twelve miles north, give or take the distance you may have drifted.[16]

Furthermore, it is relatively easy to work out the latitude by measuring the height of the sun at its highest point, the same principle that Pytheas used when he sailed north. The challenge is the longitude, knowing how far east or west you are when you are on ice that is constantly moving, with a map that has some critical errors. Nansen grumbled that the longitudes were *so narrow* that far north, roughly four nautical miles. A small miscalculation would have major consequences.

Even though Petermann Island never materialized, Nansen and Johansen had faith that eventually Franz Josef Land or some other islands farther south would appear. After all, they had no choice but to believe. I have tried to imagine what it must be like to be so isolated in the Arctic Ocean, day after day and then month after month, searching for islands that may or may not exist, heading farther and farther south without seeing any land. After five months alone on the pack ice, they finally came to a group of islands they presumed were Franz Josef Land, but could not be certain.

They decided to winter on one of the islands, and Nansen christened the place Camp Longing. They found a sturdy log that had drifted ashore and used it as the base for what Nansen described as a mud hut. It was low under the ceiling, 9.8 feet long, and just wide enough for Nansen to feel one wall against his head and the other against his feet when he lay down. The roof was made from walrus skin tied down with strips. The entrance was inspired by igloos and was so low that they had to crawl in and out, but it kept the heat in and polar bears out. They lived in the hut for 234 days, lying on a bed of uneven stones for around twenty hours a day, and surviving on meat and fat.[17] Boiled polar bear and walrus for breakfast, fried for dinner, and in spring, when they could hunt again, blood pancakes.

Nansen had a primitive soul, in the most positive sense, which enabled him to live on very little. He had learned how to survive in almost any conditions from the Inuits on Greenland. The two men had no fuel, but Nansen made blubber oil for the oil lamps. According to Johansen, he was "almost fully trained to live like a savage." Without the knowledge and skills that Nansen had acquired on Greenland, they would not have had light or the wherewithal to heat food, and would most likely have perished. Johansen wrote in his diary that they lived "more or less the same way as the wild animals that live up here in these unforgiving conditions," which to me sounds like an understatement. The wild animals have evolved to survive in the Arctic. A better comparison might be to say that they returned to the Stone Age for a while.

Børge has kayaked and walked to where the hut was on Franz Josef Land three times since 1994. The only thing that is left of the hut today is the log that they used to hold up the roof and a low pile of stones that they used to build the hut. The gray wooden log lies where it eventually fell when the hut collapsed, long after the two men had gone their way.

The Making of a Norwegian Myth

The North Pole expedition was considered a success by the Norwegian public. A record was set, and everyone survived. It stoked enthusiasm for independence. People were proud to be Norwegian, not Swedish-Norwegian, and national interest in the polar expedition became a driving force to leave the union with Sweden. The union had been established at the end of the Napoleonic War in 1814, when Sweden acquired Norway as the spoils of war. Some people argued that to break that annexation would require force, but Nansen used his fame to win Norway's freedom—a strategy that now, many years later, has become known as soft power.

Farthest North, the book about the expedition, became a best-seller, a lecture tour in Europe sold out, and stories about Nansen and Johansen were told over and over again. According to Nansen, it was only natural that Norway and the Norwegians should be superior to other nations when it came to the North Pole. We were used to snow, cold temperatures, and skiing.

The stories about Nansen and Johansen shaped more than just Norway's self-image in the run-up to the dissolution of the union. I was born sixty-seven years later in 1963 and grew up with those same stories as though they were part of my era. I also grew up believing that it was important to endure pain and be tough. I learned to deal with pain to my own advantage, but was not always so good at being tough.

One of the stories that I have heard most frequently is how they celebrated Christmas and New Year in 1895. Nansen and Johansen lay in the cramped, dark hut and dreamed they were home for Christmas. They wrote in their diaries that the two of them lay there together, in a sleeping bag made from polar bear fur, and imagined the Christmas celebrations at home with their families, the candles, warmth, and intimacy. Then Nansen interrupted himself: "But we are also celebrating in our own meagre way." He decided to wash himself for the first time since they had left the ship, *Fram*, with "a quarter of a cup of warm water, with the discarded drawers as sponge and towel. And I now feel quite another being."[18]

Wearing the same underwear all the time perhaps sounds more uncomfortable than it actually is. When you are that far north in winter, it is so cold that there are no bacteria other than those you have taken with you. When we went to the North Pole, we wore the same underwear for sixty-three days, without taking it off once. In the coldest period, we also slept in our windproof trousers, insulated trousers, thick woolen sweaters, and down jackets.

We had heard an encouraging rumor that our British competitor Fiennes was pulling 309 pounds and that he had with him an extra set of underwear, and we laughed, condescendingly. It is not really something to laugh at, but the knowledge that Ran carried more weight than us, and that we were pulling pulks, boosted us. At the same time, we suppressed another fact: Ran had more experience than us and had undoubtedly avoided many of the mistakes we had made. To save on weight, we had not taken thermos flasks, so the water we drank from metal bottles, insulated in polystyrene, froze over the course of the day. Another mistake was that neither of us had taken an extra pair of woolen mittens, which was risky. We had to take off our mittens several times a day—to eat and relieve ourselves—and when a gale was blowing, they could easily be snatched away by the wind. In addition, woolen mittens wear thin and the thumbs became so threadbare that we had to darn the holes in the evenings. We also wore windproof cotton overmittens, and had only one spare pair between us. By the end of April, they were so worn that we were constantly repairing them, and decided to share the reserve pair between us. Børge got the right hand and I got the left, and we used the material from the two worn mittens for repairs.

With 265 pounds on our pulk, "we had the feeling that we were constantly going uphill," I wrote at the time, and with 309 pounds we guessed the feeling of struggling uphill would be even greater.

HJALMAR JOHANSEN WRITES that future Christmases will be particularly sweet for the two of them, "because we have suffered and have learnt to appreciate the tiniest things."

This ability to appreciate the simplest things is a wisdom that all polar explorers share. We were very close to the North Pole when we stopped for a break and I accidentally dropped a raisin in the snow. It was not easy to get the raisins from the bag to my mouth with great mittens on both hands, but it was even harder to pick one up from the snow. I was so hungry and wanted that raisin so badly that I dropped down on all fours, lowered my head, stuck out my tongue, and flicked the raisin back into my mouth. The feeling of sheer joy when the raisin passed my lips, and the sensation of it rolling around in my mouth, of chewing it slowly, reminded me of a truth I already knew. It is all about the small things. A little tastes good, less tastes even better.[19]

Another day, when Johansen was once again exhausted, he went a step farther and quoted Fyodor Dostoyevsky: "'Man must pay for his future happiness with pain.'"[20] To remember such a line is a sign of optimism about the future. Life could be excruciating on the expedition, but when he finally got home, it would be utter bliss.

What is the meaning of life out there on the ice? I do not think either of them had any great revelations, but the answer lies in small daily miracles. "The distance between heaven and hell is so short," as Børge and I said, several times. At home, we are generally fairly balanced—things are relatively good or relatively bad—but out on the ice there is no middle ground. Life is all about either hunger, frost, and toil or satisfaction, warmth, and rest. When you go to the toilet in –40 or –60°F, you get so cold that you start to cry. You do your business just as you would at home, drop your pants to your knees, squat down, do what you need to do, then pull them up again—only as quickly as possible. Speed is of the essence, and the trick is to shave your anus before starting on a polar expedition so you do not use more toilet paper than necessary.

A few minutes later, when you have put on your skis, you head north as fast as you can, and the warmth returns. First of all in your core, then your arms and legs, and finally your fingers, toes, and nose. And it feels like heaven. Then, if the wind blows up, it feels like hell.

In the evening, once the tent is pitched, life will soon become heaven again. First the Primus has to be warmed up: The pressure in the fuel cannister has to be pumped up, and when the pressure is good, the vent is opened just enough to allow a little to drip into a small groove around the burner; then you have to strike a match on a box with extra-strong sandpaper to light the fuel. The smell of the burnt match fills the tent for a few seconds, with memories of open fires at home in Norway. The open flame burns until the fuel runs out. Another match is struck, with the same smell and fond memories, and then the actual Primus is lit. You can hear whether it is burning well or not. It often sputters and takes a while to get going because of the cold, but once the brass is even lukewarm, it burns well. The color of the flame changes from yellowish to blue. And then you know your hunger will be sated and time seems to expand.

The seconds and minutes that pass while the meat is warmed up feel meaningful. The aluminum pan is very thin to save weight, and wide at the bottom to catch as much heat as possible. You have to stir the food constantly to prevent it from burning. When the temperature is –40°F, the food can be warm at the bottom of the pan and frozen on top. Our Primus stove had been specially made and had an aluminum screen that protected both the pan and the burner, so that all the energy was used to heat what was in the pan and not the tent.[21] This meant that we used less fuel and saved on weight, but the disadvantage was that the temperature inside and outside the tent was more or less the same. As the food heats up, the smell of meat, potato, and happiness spreads through the tent. Life has never been better.

Nansen writes about the joy he felt when he managed to eke out enough thread from an old garment they no longer used to repair his outer garments, his delight at eating the breast meat of a young polar bear and their exhilaration when they came to a stretch of flat, safe ice. It is so much easier to walk on. And then the anxiety when he saw a dark cloud looming over the ice up ahead. Open water is reflected in the clouds, and can alert you to leads up to 2.5 miles away.

But perhaps the most meaningful experience for them, as it was for Børge and me, was being part of nature. It was as though our bodies did not stop at our skin but continued out into our surroundings, and the cold, wind, and sunlight became part of us. When we were moving fast, it felt as though everything was in harmony. And the togetherness, the knowledge that we were dependent on each other to survive, made us feel closer to each other.

Almost all the things that Johansen and Nansen write about are things you may have experienced farther south, but everything is more intense when you are alone on the Arctic Ocean. Their life on the ice was existentialism in practice. It is experiences like this that make it so hard for polar explorers to settle when they return home. Before long, they have to leave again.

The Christmas celebration must have been good for their relationship. On January 8, Johansen writes that Nansen suggested on New Year's Eve that they should "call each other by their Christian names." For Johansen, it was a relief that they had become closer and were less formal. Even though they shared a sleeping bag, as Børge and I did, they had continued to address each other formally. On the same day, Nansen wrote that it was his daughter Liv's birthday.[22] He was unable to sleep as he was thinking about her and his wife, and tossed and turned trying to find a position on the stones that did not hurt. This proved more or less impossible. "Today is your third birthday. You must be a big girl now." He pictured Liv dressing her up in her best dress, enjoying her day, and thought: "You probably do not miss your father." He promised that he would be there for her next birthday and tell her tales from the north. But after that promised birthday with Liv, he celebrated no more birthdays with his family.

IN HIS EULOGY for Hjalmar Johansen eighteen years later, Nansen recounted a painful memory. On March 31, 1895, Johansen tried to jump over a lead, but then it moved and widened and Johansen fell into the water. He managed to get out and they pressed on. "His frozen trousers crackled as he walked," Nansen said. Johansen asked if they could stop and pitch the tent. It was –22°F. "For goodness sake, we are not nancies," Nansen retorted. A year later, after they had spent the winter on Franz Josef Land, Johansen spoke "even less than usual for a couple of days." Then he told Nansen there was something he found hard to forget, and that was the time he had been called a nancy. "He was right, he certainly did not deserve it," Nansen said.

It was in the eulogy that he told what is perhaps the best-known story in Norwegian polar history. Nansen suddenly heard Johansen shout behind him: "Get the gun!" He spun round to see Johansen wrestling with a polar bear. He managed to keep its teeth at bay by gripping the fur at its throat. Nansen wasted a few precious seconds fumbling around, then heard Johansen shout again: "You need to get a move on, or it will be too late!" Nansen managed to shoot the bear, and Johansen escaped with a swipe over his ear from the bear's paw, which left a white stripe across his dirty face.

The story may sound a little exaggerated, but Johansen may well have been as calm as Nansen told it. When you have been out in the wilderness long enough, it feels like your surroundings are a part of you. The boundary between what is happening around you and what is happening in you is erased.

A polar bear on the trail of Erling on his journey to the North Pole.

It is therefore quite natural to keep your cool in perilous situations. Børge and I were attacked by a polar bear a little over a hundred miles from the North Pole. I heard Børge shout "*Oi!*," which I had never heard him shout before, so I looked up. About sixty-five feet away, a polar bear was coming toward us. It probably thought we were two peculiar seals. We had reached 88° 19′ N and there is not much else to eat there other than expedition teams. Polar bears usually stay farther south, where there are seals. The bear stopped when she realized we had seen her and, quick as a flash, we dug out our Magnum 44. Børge's next thought was: "*NATIONAL GEOGRAPHIC!*" He bent over, put the revolver down on the ice, and got out his camera. It was 1990, after all, and *National Geographic* was the magazine we all read. Børge dreamed of writing for them, but without good photographs that was impossible. There was no film in the camera, and it had to be put in manually. I shouted: "No, no, we have to shoot first," but Børge put the film roll in and I stood in front of him with the beautiful polar bear in the background, and a revolver in my right hand. I smiled and he got his picture. A few seconds later, the bear turned to face us directly, started to paw at the snow, and lowered her head. There was no doubt she was preparing to attack, so we jumped a few feet apart to confuse her, then she charged toward us at a furious speed. If you have a rifle it is easy to aim, but it was much harder with our revolvers. We had chosen ones with two-inch barrels to save weight. We had to wait until the bear was up close if we were going to have any chance of killing it.

Børge fired first, then I did. We kept our cool until the bear was dead in front of us. Then the fear took over, but only for a minute or two, before we once again turned back to practical matters. We paced out the distance to the bear—twenty-six feet—and then took photographs of it from all angles. This way we could show that we had shot the bear from the front. It is a sad fact that many people who claim to have shot a polar bear in self-defense, shoot it in the rump, as it is running away. I am not keen on killing animals, and we were in her territory, but we both knew it was a question of who was going to eat who for dinner, so we had little choice.

WHEN I WAS growing up, I heard stories about how ruthless polar explorers had been with their dogs. I held it against them, but that was my mistake. Even though the dogs might end up as dog food, or even, if necessary, their own, Nansen and Johansen became very fond of them. Nansen always said that Johansen had only one enemy on board *Fram* and that was the dog Svarten, who started barking and snapping whenever he saw him. The dog howled at Johansen when he sat whistling in the white watch post at the top of the middle mast, 105 feet over the water. But then, one night, Svarten was taken by a polar bear. When Johansen was asked if he was glad, he said: "No, I am sad."

"Why?"

"Well, because we didn't become friends before he died."[23]

WHEN NANSEN, JOHANSEN, and the dogs left *Fram*, they only had 180 rifle cartridges and 150 shotgun shells with them on the sleds, the reason being they had to minimize weight. The bullets were only to be used for hunting, for survival, so the men had to slit the throats of any dogs that were too exhausted to carry on. Both of them found this one of the hardest parts of the expedition. When the last surviving dogs, Kaifas and Suggen, were no longer of any use, they could not kill them with a knife. "We had become too fond of them," Johansen wrote. They agreed to sacrifice two bullets, but neither Nansen or Johansen could bear to shoot their own dog. They agreed that they should each shoot the other's dog behind a hummock. Two shots rang out.

IN SUMMER 1896, there was only one ship in that vast stretch of the Arctic.

On June 17 the same year, after fifteen months alone together, Nansen thought he heard the sound of dogs barking. Nansen was not sure if he was hearing things, but put on his skis and headed off toward the noise. Within a matter of minutes, he met the Englishman Frederick George Jackson. Jackson had applied to take part in the *Fram* expedition five years earlier, but was turned down because he was not Norwegian, and so was now leading his own expedition to Franz Josef Land, and hopefully on to the North Pole.[24]

Jackson and Nansen both described this meeting, and when I read both accounts, I get the feeling that they hoped the scene might become as famous as that between Henry Stanley and David Livingstone. According to Jackson, their first conversation was as follows:

> J: I am damned glad to see you. (And we shook hands.) Have you a ship here?
> N: No my ship is not here.
> J: How many are there of you?
> N: I have one companion at the water's edge.
> J: Aren't you Nansen?
> N: Yes I am Nansen.
> J: By Jove, I am devilish glad to see you. (And we shook hands again very heartily.)[25]

Nansen's version was shorter:
(Nansen tips his hat)
N: How do you do?
J: How do you do?
(and then they shake hands)

Nansen describes the mist that hung over them, the uneven ice that was drifting and hummocked under their feet. All around them was ice, glaciers, and the odd island.

> On the one hand, a civilized European in a chequered English suit, high rubber boots, clean-shaven and hair coiffed, with a whiff of scented soap that the savage picked up with his sharpened senses from some way off. On the other hand, a savage, dressed in filthy rages, his skin blackened from oil smoke and soot, with long, unkempt hair and a bushy beard.

As an intellectual, pioneering scientist, explorer, and artist, Nansen was, in the positive sense, the most civilized Norwegian in the world, and, for one winter, also the most primitive.

Jackson sent a man to get Johansen, who wrote that he could not talk to the first person he had seen since the previous March: "It was an Englishman, and we had to use sign language."[26] Soon, another couple of men appeared, and one of them could speak German. They could confirm for Johansen that they were on Franz Josef Land, and that the southernmost islands discovered by Payer did exist.

THREE YEARS BEFORE he died, Nansen gave a speech to the students at St Andrews University. He talked about the thirst for adventure that "propels humanity forwards on the road to knowledge" and that reflects the "spirit's mysterious need to fill all emptiness." Nansen emphasized the importance of diligence, preparation, and patience. He gave the students concrete advice: You should avoid anything that others can do as well as you, you should liberate yourself from "all necessities," find wisdom in nature, and burn all bridges behind you.[27]

It was more than three decades since he had lived up to these ideals, alongside Hjalmar Johansen. After the North Pole expedition, it was as though the desire for adventure and nature was suffocated by the expectations of wider society. One of the few instances when he expressed real happiness, after his last expedition, was when he wrote about being out in nature. He had just skied from Finse to Upsete, and on his return wrote: "It was even more wonderful than I remembered . . . Was it a hankering to go back to where the skies are high and the air is clean and life is simple?—back to solitude—silence—magnificence?"[28] There is something melancholy and painful about these questions. In the remaining years of his life, he had suppressed the childlike desire for adventure that had driven him out into nature to find pleasure in the small things that he had learned from the Inuits when he wintered with them all those years ago on Greenland.

Nansen continued to achieve great things throughout his life, and every time he succeeded, there was barely time to rest or celebrate before he had new goals in sight. The North Pole was an exception. The dream and ambition he never realized never left him. The last words he wrote, on the terrace at his home on May 13, 1930, were: "Further and further north . . ." He got no farther. His heart stopped, he fell forward and died.[29]

The Most Beautiful Love Story in the History of the North Pole

Anna Charlier's heart was buried in a small silver casket in Norra Cemetery in Stockholm at dawn on September 4, 1949. It was Charlier's final wish that her heart be cut out of her body, cremated, and taken from Devon to Sweden, to be buried with the Swedish Arctic explorer Nils Strindberg, her erstwhile fiancé. If he had still been alive, her burial day would have been his seventy-seventh birthday.

Anna Charlier was one of eleven children and lost both her parents when she was young. She was a music teacher and concert pianist. She and Nils Strindberg got engaged on October 26, 1896, eight months before he was due to leave for Svalbard. The plan was that he would join Salomon August Andrée's expedition in the hydrogen balloon *Örnen* to the North Pole, from Danskøya in the Svalbard archipelago.

In late August 2023, I sailed to Danskøya in a forty-four-foot sailing boat. There were six of us on board—Børge Ousland, Erik Engebrigtsen, Steve Torgersen, Thorleif Thorleifsson, and Håvard Tjora. We all had experience of sailing in the Arctic. From the mountains and fjords that we saw from the boat, Danskøya looked like a good choice for starting a balloon expedition to the North Pole. The island lies far to the north and the bay Andrée's expedition chose as a base is well protected from northerly, easterly, and westerly winds, but was open to southerly winds, on which they depended to fly northward. They named the harbor after Andrée's boat, the *Virgo*, and it is still called Virgohamna to this day. Older names in the vicinity reflect the lives and fates of several thousand Dutch and English whalers who hunted up there in the seventeenth century. Under a mile north of Virgohamna, you can find Likøya (Corpse Island), and to the east, Likneset (Corpse Head).

The Swedish people watched Andrée's preparations with great anticipation. A third man, Knut Frænkel, was to join Andrée and Strindberg. James Gordon Bennett Jr.'s newspaper, the *New York Herald*, offered to pay for a journalist to go with them as well, but Andrée declined. Frænkel was physically strong, a real bruiser of a man, with experience in railway construction. Strindberg had written many academic papers, but none about the Arctic, and Andrée was an engineer with forty hours' experience of flying balloons. Even though Norway and Sweden were one nation, the Norwegians were always quick to point out that Fridtjof Nansen and his men were all Norwegians. Sweden needed its own polar heroes. Andrée was frequently compared with Christopher Columbus.

No one had previously tried to fly to the North Pole in a gas balloon.[30] Andrée was a futurist and wanted to change the world. His hope was that they could cover the 620 miles to reach the Pole within forty-three hours. No one knew what the three explorers would see from the balloon. Any notions that there might be ice-free water around the Pole, or a large hole that led down into the depths of the earth, had died with August Petermann, but there might perhaps be a mountain there, or simply more ice.

Just before Andrée was set to travel to Danskøya for the second time, his mother died unexpectedly. Andrée had been estranged from his father, who died when Andrée was sixteen. Andrée's mother was the most important woman in his life and he did not want to be too close to any other woman. "I don't want to run the risk of having a wife to ask me with tears to desist from my flights," he once said, "because at that moment my affection for her, no matter how strong, would be so dead that nothing could call it to life again."[31] It seems that Andrée felt that the expedition north was a shared enterprise. He signed his last letter to her "Mamma's August," and when she then died the following day, he exclaimed that any personal pleasure in the expedition was gone. "For me, all individual interest was linked to Mamma." He said that even though he had lost his love for life, he still felt responsible for his men and for the expedition, and so would fly north. His sister-in-law said that never in her life had she seen anyone in such deep mourning.[32]

The balloon took off on July 11, 1897. It was the first time that the three men were on board a balloon together, let alone *Örnen*. They had packed all that was required should the expedition be a success: formal wear, pink silk cravats, port, champagne, and white leather gloves. They had no idea whether they would land in Canada, Alaska, or Russia, and so had both rubles and American dollars with them, as the US had acquired Alaska from Russia thirty years before.

At the moment of liftoff, a northerly wind had been blowing for some time, but then the wind changed and the members of the expedition raised their champagne glasses as Andrée gave the order: "Cut the ropes!" The tethers grounding the more than ninety-foot-high pink balloon were cut. But the balloon did not rise up into the skies. Instead it was caught by the wind and toppled over onto a padded corner post of the balloon house, the building where it had been stored. The last thing that people on the ground heard Andrée say as the balloon finally lifted was: "What the hell was that?"

None of them had done a trial flight in *Örnen* and they knew nothing about the Arctic Ocean. Strindberg's clothes were fashionable and came from European brands like Jaeger, but were highly unsuited to what awaited them. On one of the last pictures taken of him, he is smiling happily, ready to play the violin. They had dreamed of sun, dry weather, and a southerly wind. A prerequisite for reaching the North Pole was that each of these elements should be realized—all at the same time.

Prior to his first attempt in 1896, meteorologists, polar explorers, and fishermen had warned Andrée that the pack ice to the north of Svalbard was prone to hoarfrost even in summer, and frost would make the balloon heavy and unmanageable, as would fog, rain, and a prevailing northerly wind. The silk balloon had been made by seamstresses in Paris and had 8 million holes along the 45,931 feet of seam. Each stitch created a hole that the silk thread almost filled, but not quite, with the result that the hydrogen gas could escape through the holes. And it did, as well as leaking through the fabric itself. The only way to communicate with the outside world was by sending messages with pigeons, which they'd brought along.

Before leaving, Strindberg expressed both optimism and pessimism in his diary:

> Yes, I may shed a tear, when I think about the happiness which has passed and which maybe shall never again be given to me. But let me be hopeful. The balloon is varnished in any case and should be much more impenetrable than the previous year. We have the summer in front of us with its good and favorable winds and sunlight, why should our enterprise not succeed? I really completely believe in it.

Some readers have placed importance on the first part and interpreted this to mean that Strindberg realized that the journey was ill-fated. Others have placed more importance on the second half, and his optimism.

I recognize both sentiments. Before setting off on an expedition, I am always anxious. I lie awake at night and think about everything that could possibly go wrong. And there's always plenty. Before we went to the North Pole, I was unable to sleep: Had I forgotten something? Polar bear attacks, appendicitis, toothache, too little food, falling in the water, equipment breaking that cannot be repaired, temperatures of −40 or −60°F when there is no heating in the tent?

But once we are on our way, I stop thinking about almost everything. I live completely in the present, focusing on challenges at hand, and not the problems that might lie ahead.

SOON AFTER THE balloon took off, it started to spin on its own axis, only then to sink down toward the ocean. I have tried to imagine how terrible those seconds must have felt for the three men on board—when you are suddenly in free fall, when time stops, when the floor of the basket that is your home is no longer steady, and for a moment disappears under your feet and you can no longer tell up from down, or navigate from the landscape around you.

The maneuvering lines that had been constructed for navigation and to control the speed and height of the balloon had either been torn off or had fallen off and were lying down on the beach below. The three men panicked and threw nine bags of sand weighing a total of 440 pounds into the water. The sand was meant as ballast, to stop the balloon from flying too high. I think most people, if they did not know any better, would have done the same. But it was a mistake. With fewer functioning maneuvering lines and less ballast, the balloon was 1,624 pounds lighter and made an uncontrolled ascent.

Then, during the night, the wind dropped. The balloon hovered around 130 feet above ground and was enveloped in fog. In the first few days, Andrée writes that their clothes, the ropes, basket, and balloon itself were all wet, but he was not overly concerned by the dampness or lack of wind, and concluded that they were "satisfied with lying still." This seems almost incomprehensible, given the fact that time and the balloon's gas were leaking away.

Frozen atmospheric humidity creates frost—ice crystals on cold surfaces—and is every polar explorer's enemy. Anoraks, tents, Primus stoves, sleeping bags, ropes, and naturally enough, a gas balloon become less functional when the surfaces are covered in frost. And if the frost melts, this causes other problems, as everything then becomes damp before it freezes again. When we skied to the North Pole, we took off our anoraks and brushed each other free of frost every evening. It is especially difficult to brush your own back, so we brushed each other with a sturdy handbrush that we had taken with us for that very reason. It was the same type of brush that we had used during our military service to polish our shoes. Then we brushed the inside of our anoraks. Every morning, one of us brushed the frost off our sleeping bags, inside and out, while the other melted ice and made food. The ritual reinforced a good sense of togetherness.

If we had not brushed the frost away, the damp would have accumulated in the sleeping bags, and they would have quickly become heavier and less effective, like Andrée's balloon. Even though we did all this brushing, we did not manage to get rid of all the frost, and a little more ice accumulated between the inner and outer lining each night. In the middle of April, Børge and I cut open the double sleeping bag and removed all the ice that had accumulated.

Anyone who has spent time outdoors in the Arctic knows about the problems caused by frost, but the Swedes had no experience of the Arctic. Only Andrée had been in the Arctic before but had spent the time largely indoors, as part of a research project on skin pigmentation.

On the second day, Andrée writes: "We think we can well face death, having done what we have done." The temperature was just above freezing, and the balloon got damper and damper, heavier and heavier, thus reducing its load-bearing capacity. The basket started to drop down onto the ice and into the water at irregular intervals.

Courage

Andrée's expedition reminds me of Captain Walton's dream of reaching the North Pole in Mary Shelley's *Frankenstein*. Both expeditions were led by men of naive iron will, and as good as doomed to fail. But unlike Walton's situation, it was too late to turn back once they had set off.

What would have been most difficult for Andrée, Strindberg, and Frænkel? To set out on the expedition in the first place? Or to stop and turn back?

When Fridtjof Nansen made a speech for the students at St Andrews, he told them about a letter he wrote to Andrée in late spring 1897. In the letter, Nansen tells him that, in his experience, the wind generally blows from the north. This had been confirmed by meteorological observations every second hour, twenty-four hours a day, on board *Fram* from 1893 to 1896. Nansen encouraged Andrée to muster great courage to turn back for a second time: "just as he once had had the courage to turn back when he saw the conditions were unfavorable, he would also be able to demonstrate that same courage again." Andrée replied from Tromsø, on his way north, and thanked Nansen for his advice. He wrote that he was not able to show the same courage a second time. Given a choice between reason and returning home as what might be called a coward, and fortitude, continuing on with *Örnen* as a resolute hero, he chose the latter.[33]

Regardless of courage or a lack of courage, my experience is that a polar explorer's free will gradually diminishes as expectations grow. Sailing in a balloon over the Arctic Ocean was Andrée's idea once he had decided on it. But after the plan had been presented to the government in 1895 and the first attempt had failed in 1896, the idea caught the public imagination. And if King Oscar is your patron and sponsor, along with the inventor of dynamite, Alfred Nobel (who died on December 10, 1896), and the international press writes about your plans, your fellow countrymen are united in their expectation and everyone talks to you about the North Pole. I can only imagine that it would feel as though your life and freedom had been stifled by your surroundings. In a way, it was not Andrée who finally decided that they should set off on July 11, but everyone around him. Many people have said that he should have known better. But I think he did. Over the years, it has been remarked that the expedition was more or less collective suicide, or that Andrée lured the other two into his own suicide mission, but I would disagree. History has shown that nothing is absolute in the Arctic Ocean, and for a long time that was the case for Andrée, Strindberg, and Frænkel as well. As long as you are alive, there is always the possibility you might return home a hero and tell your own story about what happened.

Perhaps the venture's greatest hero was, in fact, the newly wed meteorologist Nils Ekholm (1848–1923). Following the first attempt to fly north in 1896, he realized that the balloon leaked so much hydrogen that they could not possibly fly over the Arctic Ocean. Andrée made sure to improve the balloon for their second attempt in 1897, but Ekholm remained doubtful, and he withdrew from the expedition. Ekholm was aware that he would be subject to laughter and scorn when he chose to spare his wife the worry and grief, financial problems, and loneliness, and accepted the defeat and being seen as a coward instead.

Norway is the only country in the world to broadcast detailed information about the mountain code on primetime radio and television at least once a year. Rule number eight of the Norwegian Trekking Association's Mountain Code is that there is no shame in turning back. Every now and then I ask myself if one of the reasons why I have so seldom turned back might be that I am not courageous enough to do so. The emotional burden of giving up would be greater than the possibility of trying but not succeeding. It is perhaps simply better to expose oneself to greater danger than disapproval and ridicule.

One thing that most adventurers have in common is this: We seek out challenges and danger not to dice with death, but rather the opposite. It is hugely life-affirming to experience and win in dangerous situations. It was perhaps different for the Swedes in the basket suspended under the balloon. They of course hoped they would succeed, and return home as the first people to have been to the North Pole, but it is hard to gauge what each of the three men believed would happen. "How will our journey end? Oh, how the thoughts clamour, but I must hold them at bay," Strindberg wrote in his diary on the first day. He writes about the "peculiar feeling, wonderful, indescribable!" of being underway.

Where did Andrée, Strindberg, and Frænkel, and other polar explorers get their courage? From a fear of being seen as cowards, a need for recognition, romantic dreams, foolhardiness, and/or idealism? Revenge on the naysayers? Courage goes beyond a neatly defined quality that you are born either with or without. It is something that is developed, and can be strengthened or diminished before, during, and after an expedition. Courage presupposes that you have an understanding of the possible consequences of your action. But, of course, that is not always the case. The three men had presumably not envisaged a scenario where they were marching south over the ice.[34]

The Long March Back

Örnen landed on the Arctic Ocean on the third day, July 14, closer to Danskøya in Svalbard than the North Pole. They were in the air for a total of 65 hours and 33 minutes. For 10 hours and 29 minutes they flew at the right height and maintained the desired course, the balloon hung in the air for 14 hours, and for the remaining 41 hours and 4 minutes they flew too low and on occasion the basket they were in scraped along the ice.[35]

There is no explanation in their diaries as to why they chose to land while there was still enough hydrogen to carry on flying. There is no sense of panic. The last message from *Örnen* was sent on July 13, as they floated over the ice. It arrived in Svalbard on July 15: "All is well onboard. This is our third pigeon post." It may seem wildly naive to use pigeon post to communicate with the rest of the world, but it was the only option they had. The first wireless transmission over the Atlantic Ocean wasn't until 1901. And the possibility of sending news of his heroic efforts was essential to Andrée's goal of creating headlines as a polar explorer.

Strindberg photographed Andrée and Frænkel as they stood on the ice watching the pink balloon empty of hydrogen. The strange thing about the picture is that their body language indicates optimism and determination. I assume that the pose is contrived as they both have straight backs and are looking at the balloon that is about to crumple, but there is also reason to believe that they both still thought they would survive. They both look ecstatically happy to have landed.

When I look at the picture, I am reminded of the photograph that Captain Robert F. Scott and his men took of themselves at the South Pole on January 18, 1912, when they too had a long death march ahead. The men are standing by Amundsen's tent and it is as though the air has gone out of them, with slack shoulders, hollowed chests, and eyes looking in all directions.

Both Andrée on his mission and Scott on his expressed optimism in their diaries after their photographs were taken, but the Swedes' hope is more sincere.[36] The Swedes had food for three and a half months. The Norwegians Nansen and Johansen had survived fifteen months through winter, spring, summer, autumn, winter, and spring again on their own, even though they had only taken enough food for a little more than three months. It did not seem impossible that the three Swedes would also manage, even though they knew less about polar conditions than Scott's mission and the Norwegian men.

The three men spent the first week after they landed packing a sled each with all they thought they would need for an indefinite length of time. The sleds had not been tested before they left and had to be repaired before they could be used. When they were ready, each sled weighed 440 pounds and they started to walk toward Franz Josef Land, according to the compass. Pulling 440 pounds is heavy work on flat ice, and almost impossible on pack ice. (When we skied to the North Pole, we were each pulling 265 pounds and the pulks became 2.2 pounds lighter each day, thanks to our consumption of food and fuel.)

Andrée, Frænkel, and Strindberg decided to pull one sled together, then go back and get the second, and again for the third. And so, like many polar explorers before them, they walked five times as far as they would have with lighter sleds. They eventually got rid of anything superfluous, but Frænkel kept the Swedish two-volume reference book *Jordens historia* ("World History").[37]

None of them had ever walked or even stood on pack ice. As Strindberg writes to his fiancée: "Your Nils now knows what it means to walk on polar ice." He is well aware that they will not return before the polar night falls and writes that his only fear is what Anna Charlier will think: "because now I do not mind if I must struggle and toil, as long as I eventually get home and can enjoy the pleasure of warming myself in the sun of love."

They gave up trying to reach Franz Josef Land and instead started to walk toward Sjuøyane, the most northerly islands in the Svalbard archipelago, which lay some 140 miles away, hoping to find civilization. After six weeks, they still had 123 miles to go. Most of the food was wet after the sleds had fallen in the water, and they suffered from diarrhea, constipation, cramps, open wounds, snow blindness, frostbite, exhaustion, and no doubt despondency, though it is never mentioned.

On August 13, Andrée writes in his diary that they are about to run out of meat, but later the same day Strindberg suddenly shouts: "Three bears!" The three feel a renewed sense of optimism when they see the mother and two cubs. They hide behind an ice ridge, waiting for the bears to come close enough to shoot them. But the bears keep their distance. Andrée then decides to lure them closer by pretending to be an animal that they might want to catch and eat. He gets down on all fours and starts to crawl toward them, whistling softly. The bears hesitate, but one of them is curious and comes closer, and soon enough all three were dead. Andrée writes gleefully that they now have ninety-three pounds of fresh meat, enough for another twenty-three days.

WHEN I THINK about the expedition, I sometimes wonder why Andrée and Strindberg did not write more about how hopeless it all was, and instead appear genuinely hopeful. It is paramount to hold on to hope when you are nearly broken. I try not to share negative experiences when I am in the thick of it, or swear when things get too difficult, as it can erode your spirit. As Strindberg wrote, it is a matter of holding back, of restraint. And, of course, like many other explorers, they were writing for posterity but also wanted to protect their reputation, regardless of whether they survived.

The average temperature in summer at the North Pole is around 32°F, and slightly warmer farther south in the Arctic Ocean. The ice cracks when you are pulling a sled and it moves with the wind and ocean currents. The air is often damp and visibility poor. Two of my friends, the polar explorers Cecilie Skog—who has been to the North

Pole and South Pole and has climbed Mount Everest—and Rune Gjeldnes—who has crossed both the Arctic and Antarctica via the poles—tried to get to the North Pole with skis on the ice and kayaks over the leads in summer 2011. They were in impeccable shape, were well trained, and had far better equipment than the Swedes, but chose to give up all the same. Conditions were so bad that their progress was half the speed they had estimated; they would not be able to get to the North Pole before the sun set for the winter, if they got there at all. Cecilie told me that their skis and feet were "sucked down into a thick layer of ice slush" that lay on top of the pack ice, which they could not trust to hold them as parts of it were melting. Their conclusion was that it is of course difficult to cross the Arctic Ocean when it is –5 to –60°F, but even harder when the temperature is hovering around zero. As Cecilie said: "It has never been easier to turn back on an expedition."

ANDRÉE, STRINDBERG, AND Frænkel built a shelter on the pack ice from snow and ice where they could winter. They called it *Hemmet*, Swedish for "home." A few days after they had settled in, the ice on which *Hemmet* stood cracked and they drifted off, leaving their home behind. Twelve days later, on October 5, they ran into an island that was only sketched on the map, with two alternative names: "White I or New Iceland." Andrée chose the name Hvitön, or White Island, in his diary—a highly suitable name as the island is covered by a glacier, and grayish white mist hangs over the ice for the greater part of the year. There are also more polar bears than any other animal. When they found a narrow coastal strip where it was possible to go ashore, they took the equipment, boat, sleds, and lots of polar bear and seal meat with them.

It is now known as Kvitøya (Hvitön). It is covered by a glacier and is one of the least accessible islands in the world. It is surrounded by pack ice that stretches all the way to the North Pole, nearly all year round. It is almost impossible to go ashore on the island because the glacier in most areas ends in a sixty-five-foot wall of ice, which reaches right down to the water. The glacier is also calving—ice breaks in great sheets and pieces from the edge of a glacier and falls into the water—so it is dangerous to get close.

IN EARLY SEPTEMBER 2023, the area around the island was almost free of ice. Børge and I, together with our friends Steve, Thorleif, Håvard, and Erik, sailed from Longyearbyen to Kvitøya, and after we spotted the island followed the ice wall to find a place to go ashore. On the southwest end of the island, which is now called Andréeneset, we found a coastal strip up to 1,640 feet wide between the glacier and the Arctic Ocean. This is where the Swedes prepared to spend the winter 126 years earlier. Many have wondered why they did not choose a more sheltered spot. But when you stand there on the stony shore, with a few low knolls, and look around, you realize that there is no other good alternative. To the east, there are twenty-seven miles of ice, and to the south, west, and north lies the Arctic Ocean. The terrain offers no protection against rain, sleet, wind, and snow. The Norwegian philosopher Peter Wessel Zapffe (1899–1990) described Kvitøya as "an iced tortoise of an island, a land of death and loneliness."

I am fascinated by the fact that Andrée and his companions all survived until they got to Kvitøya. Unlike Nansen, they had not learned from the Inuits how humans can survive on a desolate island in the Arctic Ocean. Almost no European had done that. Nansen and Johansen built their cabin in August, in good time for winter, whereas the Swedes were not able to until October.

Three days after they had gone ashore, Andrée wrote in his diary for the final time: They must "move a little when the weather permits."

Thomas Pynchon and the Role of the Hero

It is easy to see Andrée's expedition as a fiasco; I drew the same conclusion myself in two of my earlier books, but over time I have come to realize I was wrong. Even though we have grown up with the rule that *there is no shame in turning back*, the reality is far more complicated. For Andrée, it would clearly have been a disgrace, as for many other polar explorers.

Perhaps Andrée and Frænkel look so satisfied in the last photograph because they were happy to have flown so far north. The North Pole was the ultimate goal, but perhaps to have succeeded in flying some distance over the Arctic ice, and then to return and tell the tale, was enough. In 1897, that made you a hero. As Joseph Conrad wrote, adventurers march blindly on "to success or failure, which themselves are often undistinguishable from each other at first."

The novelist Thomas Pynchon wrote in *Gravity's Rainbow*: "What did Andrée find in the polar silence: what should we have heard?" It is an exquisitely unanswerable question. That is why the story of *Örnen* lives on.

The Husband

In 1910, Strindberg's fiancée, Anna Charlier, married the Englishman Gilbert Henry Courtenay Hawtrey. They emigrated to the US, then returned to England again in 1914. On August 6, 1930, the remains of Andrée and Strindberg were found by chance by two Norwegian trappers. The trappers were only youngsters and had never heard about the expedition. Frænkel's body was found a month later, and all three were then taken back to Sweden and given a proper burial. Anna received the letters that Strindberg had written to her, and also found out that he still had a picture of her with him in a medallion. After thirty-three years, she sent a simple wreath to his funeral, with the message: "To Nils from Anna." As the author Bea Uusma concludes in her book about the three Swedes, *The Expedition: A Love Story*: "What else could she have written? She was married to another man."

In 1947, Strindberg's niece, Ulla, visited Anna at her home in Devon. There she saw photographs of her young uncle on display, and "Anna's husband, Gilbert Hawtrey, was a wonderful person who tried to help his wife overcome her grief in every way. But to no avail. Anna could not forget."

There was a stuffed bird with open wings on the windowsill. It was the last messenger pigeon.

Hawtrey promised to honor Anna's will, and when she died, her body was cremated without her heart, then buried in Devon. And when Gilbert Hawtry died, he was buried beside her, close to where they had lived together for thirty-five years.

NOTHING IS KNOWN about how Strindberg, Andrée, and Frænkel died. There were countless ways to die on Kvitøya: polar bears, scurvy, carbon monoxide poisoning, slipping or falling down a crevasse, parasites from bear meat, vitamin A shock from bear liver, accidental gunfire, murder, suffocation if the canvas fell down over them, madness, lead poisoning from the seal on tin cans, or large doses of morphine. Håkan Jorikson, the former director of the Grenna Museum: Andrée Expedition, claims that Andrée may have taken poison with him, so that if at any point all hope were lost, they could commit suicide rather than suffer a slow death like George De Long. Or, as Andrée put it, he did not want to "drag himself on, like the *Jeannette* expedition."[38]

Peter Wessel Zapffe was a member of the team who went to Kvitøya in 1930 to collect one of the bodies. He arrived on the ship *Isbjørn* on September 5 and gave an account of the condition of the expedition base one month after the remains and equipment had first been discovered. Some of the snow and ice had melted and more of the site was revealed. Zapffe writes that it was difficult to distinguish between dirt, soil, and bodies. After an extensive and detailed excavation, they found a body: "He lay with his right hand closed under his left cheek, and appeared to be sleeping like a baby." In his jacket pocket was an envelope addressed to Strindberg, and his fiancée had written: "To be opened on your birthday on 5 September." Zapffe continues: "We looked at each other—it was the fifth of September."[39] They opened the envelope thirty-three years too late.

In the camp, Zapffe found a bottle of morphine and two tablets lying on the ice, which had not melted since 1897. The distance from Andrée's "head to the tablets was about twenty centimeters. And the bottle itself was thirty centimeters." From what I have read, the following detail has been overlooked by those who have tried to establish the causes of death: Zapffe discovered the remains of a pillow about 1.5 feet from the skulls. But there was nothing soft under Andrée's head, just two or three tins. His head was pressed so hard into the stoneface that there was not "even room to get a knife blade between them."

Zapffe raised the possibility that Andrée may have taken the morphine while standing up, then slipped down to the ground, never to wake up again. The leftovers of a meal were still in the pot. Zapffe also considered whether Frænkel or Andrée might have put morphine in the food without the other knowing, killing themselves and their companion.

THE FIRST THING I wanted to do as soon as we landed on Andréeneset, was to go to what had been the expedition's final base. I nurtured the hope that I might discover a crucial detail on the site that no one else had seen. The shoreline was covered in seaweed, boulders, small stones, driftwood, and a polar bear skull, which looked relatively fresh, gnawed clean of meat. The larger pieces of meat had been eaten by polar bears, the smaller pieces by Arctic foxes, and the remainder by insects.

When we spotted the place where the camp had been, a large polar bear was lying on a stone ledge where the men had died. The bear's fur was bloody and dirty, and she looked as though she had just tucked into a walrus or the stomach of a whale. Might she be the answer to what had killed one or more explorers?

She raised her head, watched us with curiosity for about a minute, then lay down again. We stood there, waiting for her to move, but she seemed to be both full and content. Kvitøya is the home of polar bears. Now was not the time to comb the site to see if Zapffe's theory was right, or perhaps find another one.

THE SINGULAR ENDURANCE of Andrée and other North Pole explorers is in no way unique. "It is the spirit of adventure, of daring, and the reverse of a medal," Nansen concluded.[40] In many ways it is not unlike the fanaticism born of persistence and the fight for justice, against some threat, a father or something else, rather than reason. A fanaticism that can make a person dogged in their opinions, actions, and dreams at the risk of their lives. An opinion or action that they might not otherwise have been considered in the company of those who wished them well.[41]

The Bitterest Fight in the History of the North Pole

In the history of North Pole exploration, Robert Peary fought the hardest to reach the North Pole. He led five Arctic expeditions between 1886 and 1909, and as part of these made eight attempts to get to the North Pole. He is perhaps also the only polar explorer whose life was dominated by five of the seven deadly sins—pride, wrath, greed, envy, and lust. (On the other hand, there is little evidence of gluttony and sloth in his life.)

On April 6, 1909, Peary lay down on his stomach on the ice, roughly where he believed the North Pole to be. He got out his sextant once again, raised himself up on his elbows, held the lens in front of his right eye, pointed it toward the sun, and measured its height from the horizon. He calculated that the latitude was 89° 57′ N—around three miles from the polar point. So he carried on, only to go some three hundred feet or so "beyond the Pole." The use of "beyond" plays on the idea that there is something mysterious beyond the North Pole. The expedition turned back, and soon he was, according to his own account, standing on top of the world, with his five men: four Inuits, whom he described as "hyperborean aborigines"—Ootah, Egingwah, Seegloo, and Ookeah—and the

African American Matthew Henson, whom Peary employed as his personal servant in 1889.[42]

It is one thing to reach the North Pole, and quite another to tell the world about it. Nowadays it is possible for polar explorers to call on a satellite telephone when they reach their goal, but in Peary's day the nearest telegraph station was 1,490 miles away in Indian Harbour, on Labrador.

The journey south on foot with a team of dogs and boat took five months. They arrived at the telegraph station in early September, without knowing that on August 31, the front page of the *New York Herald* had run the headline: "The North Pole Is Discovered by Dr. Frederick A. Cook." Cook had sold his story to James Gordon Bennett's newspaper. Bennett was glad he had turned down the offer to buy the same story from Peary before he left. The newspaper had made the scoop of the century.

Cook had left the US "slowly, quietly" on board a schooner named after his sponsor, John B. Bradley, in July 1907.[43] It was only in August, when Cook reached Annoatok in northwest Greenland, then known as the most northerly settlement in the world, that he announced his plans to go to the North Pole.[44]

Cook left Annoatok on February 18, the day before the first sunrise, with nine Inuits, eleven sleds, and 103 dogs. It was bitterly cold with temperatures between –60 and –40°F, but Cook writes that in the first few days he took comfort in the return of the sun. The plan was that seven of the Inuits would turn back along the way and only two, Ahwelah and Etukishook, would accompany him all the way. They had enough supplies with them for two to three months—but fourteen months were to pass before they returned.

A surefire way to earn oneself headlines is to be away considerably longer than expected, as people then assume you have died. When the news broke that Cook had reached the North Pole, it was as though he had risen from the dead. And as it turned out, that was more or less the case.

A week after the *New York Herald*'s headline news about Cook's record, the *New York Times* reserved its front page for the same news: "Peary Discovers the North Pole After Eight Trials in 23 Years." It transpired that Peary had sold his story before he left. He used the money he received from the newspaper to finance the expedition. In the event that he did not return with the news that he had conquered the North Pole, the newspaper could ask for the money back. Readers were no longer interested in the farthest-north record, and neither the *New York Times* nor the *New York Herald* wanted to pay for anything other than the news that the North Pole had finally been conquered. And now two expeditions had been the first to reach the North Pole: Cook on April 21, 1908, and Peary on April 6, 1909.

Following the article in the *New York Times*, Cook gave varying answers when confronted with Peary's claim to have reached the North Pole first. One of them was: "There is glory enough for us all." Yet the glory of being first to the North Pole cannot be shared with anyone other than one's expedition team. Peary and Cook would never be able to share the honor. Cook has since been called a *gentleman and a fraud* by those wanting to discredit him. They then often add that Peary was neither. Cook got on well with most people and was generally seen to be honest, until he claimed he had reached the North Pole. It is well documented, on the other hand, that Peary rarely behaved like a gentleman and was wont to give false accounts of his achievements.

One obvious similarity between the two is that they both were willing to risk everything, even their lives, to reach the North Pole. The two of them endured enormous suffering over a long period of time, without ever telling the full truth about the North Pole. Peary is particularly fascinating: He attempts it again and again, despite knowing what tribulations lie ahead. This shows a determination even greater than most other polar explorers, who gambled on one big expedition.

In their youth, both Cook and Peary were captivated by Elisha Kent Kane and his attempt to reach the North Pole. As a young boy, Peary cut an article about Kane out of a Sunday newspaper, which he still had when he died in 1920. Unlike Kane, who emphasized the importance of research, neither Cook nor Peary tried to disguise the fact that the North Pole was their ultimate goal. Science, in general, was deemed to be rather unmasculine in their social circles at the start of the 1900s. And it was crucial that the sleds were not weighed down by too much scientific equipment. Similarly, scientific circles began to consider the sled races as ill-suited for real research. The result was a growing mutual skepticism. For research to be valid, observations needed to be tested over time, and that is not so easy when you want to reach the North Pole as quickly as possible.

Both Cook's and Peary's lives are reminiscent of the American dream. They had humble beginnings, worked hard, and were successful. Both had a deep desire to escape their origins. Both lost their fathers to pneumonia—Peary when he was three, and Cook when he was five.

Cook's family were very poor. The last memory he had of his father was him lying in a coffin, unshaven in a dirty suit. Cook himself remained poor for the greater part of his life and died broke, despite his exploits as a polar explorer.

Peary was bullied by his classmates for being a sissy. His mother, Mary, forced her only child to dress in girlish clothes; he had a lisp and sensitive skin, so his mother insisted he wear a peaked hat to protect his face against the sun. Early in life he was convinced he had to make a name for himself: "Remember, mother," he wrote in 1887, "I must have fame."[45] And fast: "I am not entirely selfish, mother. I want my fame now while you too can enjoy it." His mother's reply would prove to be true: "such fame is dearly bought."[46] While the American explorer never became as rich as he dreamed of, he was shrewd and made good money selling things that he had paid little or nothing for, such as fox and polar bear skins, walrus bones, narwhal tusks, and three meteorites.

Peary was partly responsible for Cook's decision to become a polar explorer. In winter 1891, Cook, nine years younger, read in the newspaper about Peary's second Greenland expedition, which was due to start that summer and last a year. Peary had completed his first Arctic expedition, across Greenland, in 1886. In the article, it said that Peary wanted to have a surgeon on the expedition with him. Cook, who was a doctor, later wrote he could scarcely express the feeling that erupted in him when he read this. "It was as if a door to a prison cell had opened." He immediately sent off an application.[47]

The first time the two men met, Peary warned there was a risk that they would not survive, and they would live in such isolation that they might as well be living on another planet. I get the impression that this warning made Cook even more interested, and he was taken on as the expedition doctor and ethnographer.

For as long as he could remember, he had been drawn to danger, and he explained his fascination with a story from when he was five. Cook could not swim, but when he found himself looking at a swimming pool, he wondered how deep it was and curiosity got the better of him. He dived in where it looked deepest, "where the water was above my head, and nearly lost my life."[48] His was an extreme version of that childlike curiosity we all have.

Peary soon had reason to be glad that he decided to employ Cook. On the way to Greenland, the ship collided with the ice with such force that the iron rudder swung round and smashed Peary in the leg. Both bones in his right shin were broken just above the ankle.[49] Cook made him a splint. By late summer, Peary was able to walk with crutches, and in October, when the expedition started to make preparations for their first winter, he could stand on his own two feet.

In his book about the expedition, *Northward over the Great Ice*, Peary praised Cook's "unruffled patience and coolness in an emergency." The relationship later soured, and Peary started to hate Cook. There was disagreement as to whether Cook could publish his own articles about the expedition. Peary was adamant that he had exclusive rights, which Cook felt was unreasonable. They never went on another expedition together.

IN 1901, PEARY and his loyal companion, Matthew Henson, were away on their third expedition. No one in America had heard from the expedition since the previous summer, so Peary's family and sponsors asked Cook to go and find them.

Henson was born the year after the American Civil War ended, and slavery had been abolished. Not much is known about his early life, but his parents were persecuted by the Ku Klux Klan, his mother died when he was young, and his father a few years later. Others have said that he went to sea early and was a sailor before he got the job in the shop where he met Peary. What we do know for sure is that after he was employed by Peary, he accompanied him on every single expedition and became more and more indispensable for Peary, not only as a servant but as a fellow polar explorer. Henson chose to live with the Inuits, rather than the Americans, for long periods. He was no doubt relieved finally to find people who did not look down on him because of the color of his skin. He also learned to mush from the Inuits.

Mural of Matthew Alexander Henson, who accompanied Robert Peary, planting the American flag at the North Pole. Austin Mecklem painted this at the Recorder of Deeds building, in Washington, DC.

Peary was always keen to show that he was exceptionally tough. He claimed that he could deal with the cold, as long as he had good clothes and enough food, but "the darkness, the months-long winter nights! That is different."[50] In the book about his third expedition, Peary writes that on February 18, 1901, he could barely stand on his feet due to frostbite. He was in Fort Conger, Adolphus Greely's former base. When Henson helped him to pull off his boots, the end joint on two or three of Peary's toes came off with the inner lining. Henson asked why he had not mentioned the frostbite, as gangrene was setting in, and Peary replied: "There's no time to pamper sick men on the trail," and that a few toes were a small sacrifice to "achieve the Pole." Peary does not mention his toes again until March 13, when he writes briefly: "The final amputation of my toes was performed." He claims that he had to amputate eight toes and only had his little toes left. Henson said that all bar one fell off.

On their way to the base, Peary wrote: "I shall find a way or make one." These words were later inscribed on his gravestone, which is shaped like a globe. The quote is normally ascribed to the Stoic Seneca,[51] but Hercules is also quoted as saying it in Greek legends, as is Hannibal, when people told him he could not cross the Alps with elephants.[52]

When Cook finally found Peary, not only had his toes been amputated, but Cook felt that his colleague's health had suffered seriously as a result. He was depressed and anemic, and feared his chances of reaching the North Pole were minimal now that he only had two toes. Peary says nothing about Cook's visit in his book about the expedition, but does mention his feet again in connection with the time spent "breaking in the tendons and muscles of my feet to their new relations."

Cook's conclusion was that Peary was finished as an explorer, which perhaps came as a relief. But Cook was unfortunately rash enough to tell Peary that he would never be able to use snowshoes or skis again with only two toes and eight painful stumps. Peary never walked normally again, but, in Henson's words, did manage to master a "peculiar stride-like gait," which proved effective enough. In addition to his greed for fame and fortune, Peary was also driven by a desire for revenge, as most polar explorers are. To avenge himself on the peers who had bullied him when they were young, on people who were more successful, and a fate that he felt was unfair.

I will show them has been the secret creed of many a polar explorer.

COOK'S EXPEDITION TO find out what had happened to Peary reminds me of Charles Marlow's boat journey into the interior of Africa to find Kurtz, in Joseph Conrad's novel *Heart of Darkness.* No one had heard from Peary or Kurtz, and Cook and Marlow were sent into what was ostensibly the unknown to establish their fates. There are also similarities between Peary's life with the Inuits and Kurtz's life as the commander of a trading post far upriver in Congo. Both are, for all intents and purposes, rulers of their own kingdoms. They are dependent on the local population and make sure that the locals are dependent on them. Like Kurtz, Peary had control over all goods that arrived at Melville Bay, in Greenland, he talked about the Inuits as *his*, and those who did not work for him were excluded from the community's growing cash economy.

When Marlow finds Kurtz, cut off from humanizing influence, loneliness and alienation have broken him. He appears as a person who acts on his most ruthless instincts and resembles more of a monster than a human. Peary, in his isolation, would risk his life to get to the North Pole with some of the same Inuits, and was kinder than Kurtz.

The explorer talked about the Inuits like children and said they "should be treated as such." He claimed their feeling "for me is one of gratitude and confidence." In Peary's eyes, they were an inferior race, as was Henson, whom Peary insisted should call him *sir.* Whenever he arrived at Melville Bay by boat from the US, he shouted for his own amusement to the shore: "*Tikeqihunga*," which means "I am coming now." According to one eyewitness, he opened a barrel of biscuits for the first Inuits to come aboard, and they ate as many as they could and stuffed even more in their pockets. The barrel was then later taken ashore and the contents strewn on the beach.

Men, women, and children threw themselves over the biscuits, while Peary stood and watched.[53]

In 1897, Peary traveled to Melville Bay, not to winter there or make an attempt on the North Pole, but to collect the meteorites that William Parry had heard of in 1818. These were a source of iron for the Inuits, and Peary had found them three years previously. In addition, the anthropology department of the American Museum of Natural History in New York had asked him to take back a middle-aged Inuit as well.

Late in the summer, Peary returned to New York with six Inuits: father and son Qusik and Minik, the shaman Atangan and her husband, Nuktaq, a hunter, and their adopted daughter, Aviaq, and her fiancé, Uisaakassak. None of them had been forced to leave, but Peary had lured them away with talk of warm homes and sunny climes, guns, knives, and needles, and had promised them that they could go back the following year. Even before they arrived in New York, 20,000 people had bought tickets to see them in their traditional costumes, as well as the meteorites. All the profits went to Peary. The six Inuits were housed in the basement of the museum, and after some time as research objects, four of them died from tuberculosis. Minik's father, Qusik, was one of them, and his skeleton was stripped and cleaned by museum staff and then exhibited alongside stuffed animals from North Greenland, where they were all born and had grown up. The museum organized a bogus funeral for Qusik, so that Minik would think that his father had been buried in a worthy manner and not become part of the exhibition that he later went to see.[54]

The meteorites were exhibited elsewhere in the museum. The largest weighed thirty-one tons. Officially, they were allegedly a present from Peary but, in reality, they were only on loan until the museum had paid for them in full. Peary knew how unique the meteorites were so they became his most profitable deal.

TEMPTING AS IT might be to say that Kurtz and Peary were children of their time, I am not sure. As far as I know, no other polar explorer has treated the Inuits as badly as Peary did at times. But as the Canadian author Kenn Harper writes in the book *Give Me My Father's Body*, some Inuits also "genuinely liked the man." Harper has lived in the Arctic for more than half his life, speaks Inuktitut, the Inuit language, and spent much of his adult life researching and sharing Minik's story.

The explorer dressed like an Inuit, mushed dog teams like them, learned to build igloos, and knew enough of the language to get by. Thanks to him, the settlement in Melville Bay became less isolated, men and women got work, food was imported, and they had access to better tools for fishing and hunting. Today it might be natural to question if that was in fact progress, but as Harper points out, the Inuits had already been trading with whalers for a long time. Whaling had been so successful in the area that the number of whales had fallen dramatically, and the whalers had therefore stopped coming there. There were no trading stations nearby, and the Inuits had come to depend on things such as good knives and guns to secure food, and thin, robust needles.

One of the things that I admire about Peary is that he mucked in and helped with all manner of tasks on the way to the North Pole. Many other expedition leaders chose otherwise and left the heavy work to others.

Peary disliked having anyone he could not control in the vicinity. On the American's third expedition, the Norwegian polar explorer

Otto Sverdrup wintered there for four years in the same area. Sverdrup's expedition was there to explore North Greenland and the islands to the west, but Peary was convinced that Sverdrup was lying and was actually trying to reach the North Pole. And he hated him for it. It was as though he believed he had the God-given right to get there first. Most polar explorers have rather strained relations with other polar explorers. You have to risk so much to achieve your goal, which is often to be the first in some way, that anyone else trying to do the same is automatically a threat.

In the four years that the two were living in the same area, the American tried as much as possible to avoid Sverdrup, and the Norwegian tried as much as possible to make contact. They met just once, and only by accident. Sverdrup was out on a dogsled and had stopped on the ice to warm himself with a cup of coffee. Peary came by on his own dogsled. Sverdrup took off a mitten and held his hand out, but Peary kept his mittens on when they shook hands. The Norwegian offered him a cup of coffee, but Peary declined. Their meeting was so brief Sverdrup "barely had time to take his mittens off."[55]

When Peary came home from the expedition to America, he claimed to have discovered Axel Heiberg Island, which was not true. The American Geographical Society and Royal Geographical Society believed him, until it was proved that Sverdrup had not only discovered, but also mapped the uninhabited island.

PEARY'S ATTITUDE TO women was no doubt compounded by the fact that he, like Kurtz, spent too long without the company of people who might correct his behavior. He maintained that in a climate such as the Arctic, a man had the right to have sex with local women. He was certainly not alone in thinking this, as children were born and abandoned by their fathers, and sexual diseases were rife. But Peary's thoughts on a polar explorer's desire went one step further: "Let white men take with them native wives, then from that union may spring a race combining the hardiness of the mothers with the intelligence and energy of the fathers. Such a race would surely reach the Pole if their fathers did not succeed in doing it."[56]

PEARY'S STORY SHADOWS that of Captain Hatteras in Jules Verne's novel of the same name: Both wanted to go to the North Pole, and both claimed to have discovered new territory in the Arctic Ocean. Hatteras named the island he discovered New America forty-five years before Robert Peary claimed to have discovered Crocker Land. The islands share the same location, to the northwest of Greenland. Peary believed that Crocker Land could be the world's last undiscovered continent. He estimated it to be "nearly half a million square statute miles," including islands, shallows, and land, which is roughly three times the size of Norway.[57]

Ice formations in the Arctic Ocean can often look like land formations when the light is low, and it can be hard to gauge the distance between yourself and what you see. Everything appears white, and it is hard to even differentiate between the snow beneath your feet and the air around you. The only things that stand out are brightly colored sleds, skis and ski poles. Børge and I used to joke that it feels a bit like skiing inside a milk carton. A dark ice formation a couple hundred feet ahead can look like a mountain a thousand feet away. But to see a continent on the ice that is not there seems absurd.

In his expedition diary, Peary did not mention the discovery of Crocker Land; nor did he say anything immediately on his return, not even to his patron George Crocker, after whom he had allegedly named the island.[58] It was only six months later that he "rediscovered his own discovery," as was said at the time. The territory was drawn on new maps, until an expedition to Crocker Land in 1913 found that there was no Crocker Land, only more pack ice.

BOTH COOK AND Peary claimed that getting to the North Pole was a calling greater than themselves. Cook writes about the pull in his book, *My Attainment of the Pole.* He believes the need "to invade the unknown" was buried within him even before he was born and he compares himself with Icarus: "And I felt the indomitable, swift surge of their awful, goading determination within me—to subdue the forces of nature, to cover as Icarus did the air those icy spaces, to reach the silver-shining vacantness which men called the North Pole." I think "the silver-shining vacantness" is one of the most poetic and precise descriptions of the North Pole that I have ever read.[59]

Peary's compulsion to be first to the North Pole had a divine dimension. He describes his calling in a letter to his mother: "In the irrevocable book that I have been selected for this work and shall be upheld or carried safely and successfully through." Three years before he allegedly reached the North Pole, he uses nothing less than the Earth Goddess Gaia to explain his ambition. He wanted to travel toward the Goddess and get as close to her heart as possible. And the closest was "up there in that borderland between this world and interstellar space which we call the Arctic Regions."[60]

On his final attempt to reach the North Pole, Peary is said to have carried a poem with him. "The Frozen Grail" by Elsa Barker. It is a poem about Peary the hero—judging from his narcissism, he is unlikely to have taken a poem with him that described any other subject than himself. She describes Peary as an Arctic Galahad, one of the knights of King Arthur's Round Table. According to the legend, Galahad was victorious in a duel with his own father before he found the Holy Grail. The North Pole is depicted as a heavenly kingdom, where "the immaculate Virgin of the North" guards the Holy Grail. The poem points out that brawn is required, as the virgin "must be taken only by violence." Barker asks if he who conquers the world must also renounce the world, and answers that these men have already done so. The poem was published in the *New York Times* both before and after the expedition. All ends well, according to the poet. When Peary reaches his destination, he will "lift his warm lips to the frozen Grail."

The Fight to Be Believed

It is not easy to be sure if Cook or Peary or neither was telling the truth. It is even harder to believe that they were both telling the truth. Whatever the case, neither Cook nor Peary could prove that they had reached the polar point.

Providing evidence should have been relatively uncomplicated. For example, diary entries with the height of the sun measured with a sextant day after day, together with the calculations of latitude and longitude, would be ample proof.[61]

Peary held a vital card that almost made up for the lack of evidence. The expedition had the support of the National Geographic Society, the US president, Theodore Roosevelt, the *New York Times*, and several of the richest men in the States. They all had a vested interest in Peary's success, as well as a financial stake. In his introduction to Peary's book about the expedition, Roosevelt underscores that, by reaching the North Pole, Peary "has made all dwellers in the civilized world his debtors."[62] The president was concerned that the lives of Americans had become too civilized and that social ideals had become too feminine. He believed the nation needed masculine role models like Peary, who left civilization behind to challenge the dark, the cold, and themselves.

It is hard to win against the wealthy elite of America, and Cook accepted that there was little likelihood that he would be recognized as the person who discovered the North Pole. "I had not the money nor the nature to fight in this kind of battle—so I withdrew." But he continued to claim this right for the rest of his life. *My Attainment of the Pole* is a bitter attack on Peary and his supporters. In the introduction, Cook promises the reader that he will "use the knife. I shall tell the truth even though it hurts." Cook's truth was that Peary's supporters were corrupt, and that Peary was a liar "with the hand of a buccaneer and the heart of a hypocrite."[63] He had nearly killed Cook by stealing food from a depot, something that he felt justified death by hanging for Peary.

Peary took a different tone when he told his story to the newspapers. To the surprise of many, Peary asked the poet Elsa Barker to write the first article about his expedition. It was relatively usual for polar explorers to let others write their story, and equally usual to hide the fact, but to let a woman write on his behalf was a novelty. My impression is that Peary was not even vaguely interested in the equality of the sexes, and his decision was influenced by the fact that women had become consumers. They bought newspapers and books and often made up half the audience at his lectures. A woman had written to Peary in 1907, having heard one of his lectures, and explained that his book could be improved by avoiding sentences such as "I was more than glad when we stepped on firm ice." She recommended that he spend more time on "that soul-stirring incident crossing the young ice on snow-shoes."[64] It was time for a new style of telling.

Barker worked for *Hamptons* magazine, and the magazine bought the rights to the whole story before the book was published. She wrote the story, but by this point the general public were already bored of the fight between Cook and Peary, and the magazine failed to sell as well as expected. Peary was disappointed, and when the time then came to choose an author for the book about the expedition, he once again chose a man. Perhaps he was also tired of Barker's thorough approach. She had repeatedly asked him for more information about the way back from the North Pole, a part of the expedition he had not documented particularly well.

AFTER PEARY'S DEATH, his family and his benefactor, the National Geographic Society, controlled the polar explorer's archive. The family chose to give only very limited access to researchers. However, in 1984, after seventy-five years, both parties decided that it was time for a change, and the North Pole explorer Wally Herbert was commissioned by the National Geographic Society to study the contents of the entire archive.[65] Naturally enough, Herbert started with Peary's diary from 1909 and went straight to April 6, the day he supposedly reached the North Pole. But the page was empty. However, a piece of loose paper had been inserted, on which it said: "The Pole at last!!! The prize for three centuries, my dream & ambition for 23 years. *Mine* at last." Nothing had been written on April 7, the second day at the North Pole, nor April 8, which was also blank. Herbert turned back and started to read from the beginning. The more he read, the less he believed Peary's claim, until eventually he rejected it completely. The diary had very few stains and did not look as though it had been on a long journey on a sled. Perhaps it was a facsimile of the original? As a reader pointed out, given how much pemmican Peary had eaten on the expedition, there would surely at least be grease marks on the original.

One question that remained unanswered is why Peary organized his expeditions in such a way that only he could calculate and work out their position with the help of a sextant. A common conclusion is that he did this to avoid witnesses, but it may also have been because he wanted to be the only white man to the North Pole.

Herbert concluded that Peary was bluffing about his positions in both 1906 and 1909. In his diary from 1906, the only day that he missed is April 21, the day the record was set.

The British explorer persisted. He wanted to get to the bottom of who was first and so studied all of Cook's documentation as well. Like Peary, he had no witnesses, and the Inuits who were with him could not use a sextant. Herbert writes that in theory Cook, like Peary, may have gotten that far north, but the lack of documentation might indicate that he chose what Herbert elegantly calls "the easier route to fame."

Herbert also talked to Talilanguaq, one of Peary's grandchildren. No one in the family had seen their grandfather since 1909, when Henson and Peary were on their way south to telegraph. They stopped to see their sons and their mothers, Ahlakaceena and Anaukaqto, who had given birth on board Peary's trapped ship three years earlier on his fourth expedition. Peary had another son from the third expedition. Henson and Peary soon traveled on and their sons grew up without a father, as they themselves had done. Talilanguaq was a skilled musher and believed that his grandfather had reached the polar point, but not Cook, as going to the North Pole with three men, dogs, and two sleds was "to commit suicide."

A public dispute between two North Pole explorers about how far north they had gone was something new. Until 1909, polar explorers were believed when they said how far they had gotten (with a few exceptions). Had only one of Peary or Cook returned and told his story, he would in all likelihood have been accepted as the first man to reach the North Pole. Certainly until Herbert read the diaries.

The Danish polar explorer Knud Rasmussen gave credence to Cook to start with. Inuits who knew Etukishook and Ahwelah told him that they had asked Cook to turn around repeatedly, as they had the feeling that they were very, very far from land and did not believe that they would get home alive if they carried on north. They had also said that the pack ice at the North Pole looked just like the ice farther south, which is true, and they described how Cook had jumped and danced like a shaman when he realized he was in the vicinity of the polar point.[66] Rasmussen later changed his mind when he discovered the scant documentation that Cook could provide, and he called what he saw a childish attempt at cheating.

"Who Do You Think Reached the North Pole First?"

Some of the stories told about North Pole expeditions seem almost too bizarre to be believed, but then, in my experience, nothing that happens on an expedition is too bizarre to be believed. The National Geographic Society accepts that Peary did not actually reach the polar point, but that he *believed* he did. If he was thirty miles away, I doubt he would have believed he was at the North Pole, but he may have believed he was in the vicinity. The Explorers Club, another of Peary's original sponsors, still credits the expedition as the first to reach the Pole, "or at least come close." The *New York Times* has been more frank and in 1998 wrote: "Today most historians believe both explorers were lying."

The answer also depends on what is meant by reaching the North Pole, and the definition of the polar point changes. Nowadays, with GPS, you should be within three feet of the polar point. It is not easy to be at the polar point for long, as even if you are standing still, the ice is moving. Most people agree that the North Pole in 1909 should be afforded a wider radius than today. One suggestion is six miles. I think that is rather harsh as the natural margin of error may have been greater. A sextant is designed to measure the correct position, but more than a hundred years ago clocks were less exact. The clouds generally hang heavy over the Arctic Ocean and sometimes it is difficult to see the sun to measure its height above the horizon. A sextant that has been shaken on a sled moving over uneven ice for weeks may well give abnormal deviations. And in the time between taking one measurement and the next, you may have drifted so that the one position does not accord with the other. It is far simpler at the South Pole, as you are standing on ice that rests on a continent.

In the future, with increasingly accurate measuring instruments, it may be possible to define the North Pole down to the nearest centimeter, and you will have to stand there to document that you have reached the North Pole. When we got to the top of the world, the polar point was in an open-water channel, and it was only several hours after we had pitched our tent that we drifted over the point itself. We had reached our goal, but had no way of knowing at that moment if we were the first to reach the North Pole without depots, engines, or dogs, and had beaten the British. If they had gotten there the day before, they would have drifted out of sight. We had no satellite telephone, only a VHF radio with enough range to talk with any airplanes in the vicinity, and a transmitter that sent our position and three simple preprogrammed messages ("distress," "great distress," and "we have arrived") out to the world.

The first morning we were there and rationed out the little food we had left, while waiting for the plane that was coming to collect us, we suddenly heard engine noises. We were about to prepare a landing strip on the ice, when I turned to face the noise and saw a large white propeller plane flying low, less than four hundred feet away. The plane had "US NAVY" written on the side, but it had no skis. It turned out to be a spy plane that dropped probes into the open leads in the Arctic Ocean to localize Russian submarines under the ice. Børge turned on the VHF radio (we had not taken a bigger radio because of the weight). I stamped out the word "FOOD" in big letters with my feet in the snow. Børge made contact and answered in a strong Norwegian accent when the American voice asked who we were: "We are just two Norwegians that have walked to the North Pole." The next question was: "Do you need food?"

The Americans dropped a container of food and another with reading material, which was limited to *Penthouse* and *National Enquirer*. Pictures of naked women were redundant when we were so exhausted, so I read every article in the gossip magazine, *National Enquirer*, instead. We were more interested in the food. Even though we had been eating 5,850 calories a day, our fat and muscle mass was utterly depleted. We had started to ration our daily intake as soon as we got to the North Pole, as we had no idea when we would be picked up. We divided the food from the plane equally and laid it out on our sleeping mats, sticking to the method we had used every day for breakfast and dinner, where one of us divided up the food and the other chose his portion. I remember desserts were shared the same way at home when I was growing up.

It was a memorable meal. This is what I wrote about it:

> I was about to bolt it all down, but Børge suggested that we wait a little, not eat straight away. Observe the food in silence. Count to ten in our heads and then eat. Show shared restraint. Remind each other that contentment often involves sacrifice. I have seldom felt as rich as I did then. It felt strange to wait, but the food tasted all the better for it.[67]

WE CAN BE fairly certain that Peary did get as far as 87° 47′ N. The position was measured by Bob Bartlett, the only person other than Peary on the expedition who could use a sextant, and it was his last measurement before he was sent back. Before the expedition set off for the north, Bartlett told his father, who had been a seal hunter in the Arctic for most of his life: "On the next voyage I'm going to the North Pole."[68]

Until they got to 87° 47′ N, Bartlett had gone ahead of the others to prepare the route through the ice, and so made a key contribution to the expedition. Bartlett believed that Peary had promised he would go all the way to the North Pole, and that he was betrayed. The morning before he was forced to go south again, he got up before all the others, took his equipment, and walked some way north by himself and, by his own admission, shed a few tears: "perhaps I cried a little."[69]

An amusing detail is that even Peary and Henson could not agree on who got to the North Pole first. Henson said he did. On the day they reached the North Pole, he was walking ahead of the others to work out the best route. Once they were at the polar point, Henson said he had gotten there first, and stood by this claim for the rest of his life.[70] Peary denied this, of course, and to this day no one knows.

As far as I'm aware, no one ever asked the four Inuits about it. For them, it was a job, and they would not be celebrated on their return. I am not sure that any Inuit at the time would see the point of going to the North Pole. For a long time, it was only European and American men who were obsessed with the idea. In many ways the situation was similar to that of the Sherpas and Western men who wanted to climb Mount Everest in the early days. When he was a boy Jamling Tenzing Norgay, son of Tenzing Norgay, asked his father to help him join an Everest expedition. His father replied: "I climbed Mount Everest so my children wouldn't have to and could go to school instead."[71]

Neither Peary nor Henson denies in their books that the four Inuits were crucial to the expedition. On the last day, they saved the lives of both Americans when Peary and Henson fell in the water. Henson describes the incidents "as part of the day's work."[72]

MORE THAN A hundred years later, perhaps the march to and from the polar point is more interesting than whether they actually got there or not. Irrespective of what was promised, and how much of it is made up, Cook, Peary, and the other seven men are towering figures.

But Cook, Etukishook, and Ahwelah's most impressive feat was in 1908–9, on their return from the North Pole. They were forced to winter there, like Fridtjof Nansen and Hjalmar Johansen. The sun disappeared on November 3 and did not rise again until February 11. "We were therefore doomed to hibernate in our underground den for at least a hundred double nights before the dawn of a new day opened our eyes." Like Johansen and Nansen, the three men returned to the Stone Age and survived on meat and blubber from seals and polar bears they had killed before they settled. The sleeping platform they shared was made from stone and life gave "no excuse for cheer. Insanity, abject madness, could only be avoided by busy hands and long sleep." Naturally, they did not have enough matchsticks for an extra winter, and so set up a rotation of six-hour shifts to make sure the lamp did not go out and to scare off the polar bears and "to force an interest in a blank life." It was pure luck that Johansen and Nansen were rescued so soon after winter. Cook, Etukishook, and Ahwelah did not have luck on their side and had to walk all the way to Annoatok. They ran out of food on the way, and had to eat two things not normally considered edible: one was candle wax melted in warm water, the other was the hide of a boiled walrus.[73]

The fact that the three men survived for fourteen months, before finally returning home, is one of the greatest feats in polar history.

THE HEROIC ERA in the Arctic did not end overnight, but the conflict between Peary and Cook marked the beginning of its end. Most people believed the North Pole had been conquered once and for all, and the race was finally over.

Interest in the two American explorers dwindled in the buildup to the First World War. And when the war broke out and millions of young men were being killed and maimed in the trenches, the explorers' drawn-out personal feud and attempts to reach the North Pole must have seemed absurd.

Instead, a new world was being born.

7

THE MECHANICAL AGE

Meteorologist Finn Malmgren and others in the cabin of the airship Norge *in 1926, which overflew the North Pole. He would later die on a second journey there.*

Flying to the North Pole

Roald Amundsen's home has stood untouched since he left the house for the last time on June 14, 1928. The property, which is called Uranienborg, lies just outside Oslo. The only toilet in the house is just inside the front door, and on the toilet seat is a note that says "BLOCKED!!!" The message to say the toilet should not be used is perhaps the last thing he ever wrote. In the living room, there is a copy of a painting of the South Pole explorer "Titus" Oates as he leaves Captain Robert Scott's tent. The entry in Scott's diary for March 16, 1912, says that a storm was blowing and the temperature was down to −40°F. As Oates crawled out of the tent, he famously said these last words: "I am just going outside and may be some time." Amundsen loved stories like that.

The shelves in his study upstairs are crammed with leather-bound books about the poles, encyclopedias, and Bibles, protected by glass cabinet doors. The front cover of the magazine *The Frontier* on his bedside table shows a cowboy in a blue shirt and orange neckerchief holding a six-shooter. Amundsen was an avid reader of stories about the Wild West and left behind a large collection of magazines. Jack London's *Call of the Wild* was his favorite book, and is still to be found in his home.

Two metal springs with wooden handles are attached to the study doorway, for strength training. Amundsen would stand there with a straight back and work his biceps and triceps. He was proud of his physique and said that a military doctor had been so impressed by the future polar explorer's well-trained body that he asked the other doctors to come and have a look at him. When Amundsen sat at his desk and looked out the window, he could see over Bunnefjorden and Oslo Fjord to the hill where Fridtjof Nansen's home lay. Polhøgda, with all its pillars and towers, is where Nansen lived for the last thirty years of his life. Nansen was more than a role model and mentor for Amundsen. Early on in life, he deified Nansen, but then they became competitors of sorts—to be the greatest polar explorer of all time, though this was never said explicitly.

Nansen was a talented man in many fields, whereas Amundsen was good at only one thing: planning and carrying out polar expeditions. If Nansen had not been interested in politics, humanitarian work, and writing, he could well have been the greatest polar explorer in history. But he chose to spread his talents, leaving the way open for Amundsen to claim the title.

The North Pole was the first pole that both Nansen and Amundsen set their sights on, but neither of them wanted to be the second person to get there. So when Cook and Peary claimed to have reached the Pole on foot in 1909, an expedition across the ice became pointless. Amundsen was one of the people who immediately supported Cook's claim. The American had become a friend for life, after they'd spent two winters in a row together in Antarctica, from 1887 to 1889. Amundsen was loyal to those who were loyal to him, even when the world turned against his friend.

As a polar explorer, Amundsen was both old-fashioned and modern. In 1907, he looked into the possibility of using tame polar bears to pull sleds instead of dogs. During a visit to Hamburg Zoo, he was convinced, briefly, that a *trained* polar bear was the most cooperative animal on four legs. "They are more amiable than dogs," he said.

Amundsen followed any innovations that might improve the possibility of reaching new destinations by flying. From the start of the century, he watched the development of kites, hot-air ballons, flying boats, and airships with keen interest. And in the years immediately after the First World War, he was gradually convinced that flying boats were now so advanced that they could reach the North Pole, so in 1924 he made up his mind.

Fifteen years after Cook and Peary had purportedly reached the North Pole, he saw the opportunity for a new race and a new record: to fly to the North Pole. There were already rumors of possible competitors. It is no longer clear how true these rumors were, but Amundsen felt, as so often in his life, that time was of the essence.

The plan to fly to the North Pole heralded a new era for explorers: the mechanical era. Dogs, wooden sleds, and resilience were replaced by engines, ice-cold metal, and the expertise of engineers. An expedition to the North Pole would no longer take years, and would involve far less cold and frostbite, but the risk was even greater.

This was the era Joseph Conrad named *Geography triumphant*. Conrad maintained that history had finally entered this third era, that of the modern world: a time that no longer fostered heroes who walked or sailed into the unknown to chart new worlds. Explorers could now read maps made by others and follow already trodden paths—more like a tourist (though Conrad did not use that word in his essay).

And Conrad was right: After more than 400 years, the world was more or less mapped, with the exception of large areas of the Arctic Ocean. In the time since the article was published, explorers and other tourists have become more and more alike. Nowadays, expeditions are organized whereby you can fly almost all the way to the North Pole and then walk the last few miles. I see no wrong in that, other than the question of sustainability, but it is, as the author predicted, a more civilized experience.

The flaw in Conrad's conclusion is that he seems to think that progress has slowed. As with many people who have lived a long life—the article was published the year before he died at the age of seventy-seven—he believes that the world of yesteryear was more exciting. Two momentous eras in the history of exploration—*Geography fabulous* and *Geography militant*—were over, and the future could never be as fascinating. But every epoch in human history has its own stories of apparently unconquerable natural forces. Unexpected challenges always appear. New heroes who resemble those of old are born, and even though the goals may differ, the dreams, wonder, and hunger for adventure are the same.

The path of wonder and adventure was already well trodden before Conrad's *Geography fabulous*. It stretches from the first people who tried to visualize the imaginary North Pole by studying the celestial North Pole, to prehistoric times when men and women walked and sailed north, west, and east, and then in all directions to explore and create new lives, to Pytheas and the seafarers of the fifteenth and sixteenth centuries, and on to the polar explorers from the 1600s until the day that Roald Amundsen decided to fly to the North Pole.

The path then continues past the North Pole to the moon and on into space, forged by our desire to understand who we are, where we come from, and what might happen in the future. As Amundsen wrote in 1926, two years before he disappeared: "I am glad I will not live much longer. Then the moon would be the only chance."

An Apparent Miracle

It seems to me that the only constant when planning North Pole expeditions is the mounting costs. In October 1924, Amundsen faced not one but two challenges when it came to money.

The first was to finance the expedition itself, and the second was to avoid personal financial ruin. Like many other polar explorers, he used his earnings from lecture tours, book sales, and other income to pay off any debts from previous expeditions. And he was also willing to bank on possible future income as a guarantee for expenses in connection with the new expedition.

The Norwegian and international press, who had built Amundsen up to be a great hero, "the man who discovered the Northwest Passage" and "the man who conquered the South Pole," now wrote about his financial problems, implying that he was unpredictable and chaotic. It was a dramatic fall from grace. His creditors openly started to prepare for his eviction and the sale of his home in Uranienborg, and the likelihood of securing funds for a new expedition was as good as nothing. But when everything looked utterly forsaken, two things happened that must have felt like a miracle.

Two loyal friends and supporters with Norwegian roots, one in Chicago and the other in Buenos Aires, said that they were willing to buy his house when the creditors put it up for sale, and to make him the legal occupant. They both knew about his plans to fly to the North Pole, so in order to avoid their friend having to invest all that he had and more, yet again, they wanted Amundsen to be able to live in the house, but not own it.

The other miracle happened at the same time that his friends came to the rescue. Amundsen was on a lecture tour in the US, and had just arrived in New York. He hoped to find sponsors for his expedition while he was there. When not living in a tent or on board one of his expedition ships, *Gjøa*, *Fram*, and *Maud*, Amundsen enjoyed a good dose of luxury. From his room in the Waldorf Astoria, he writes that he feels "the future had closed solidly against me, and that my career as an explorer had come to an inglorious end." As he sat there alone in a chair, ruing the fact that he lived in a time when courage, willpower, and belief were no longer enough, the telephone rang.

Amundsen assumed it was a creditor and considered not answering, but, fortunately for polar history, changed his mind.

A stranger was asking for him in reception. The stranger introduced himself and explained that he was an amateur when it came to polar expeditions, but he could raise the money. As Amundsen later said: "I'm sure I don't need to tell you, but I immediately invited him up."[1]

The man was Lincoln Ellsworth (1880–1951). He shared President Theodore Roosevelt's concern, voiced by the former president in the foreword to Robert Peary's book in 1909: The lives of people on the east coast of the US were too civilized and too removed from nature. Like Roosevelt, Ellsworth glorified life in the western part of his country, where men slept under the open sky and were responsible only to themselves, and, more often than not, the hero was a cowboy who could survive with "rope and revolver."[2] For Ellsworth, as a wealthy man with a thirst for adventure, and a deep desire to discover new countries and contribute to cartography and geography, the Norwegian polar explorer fit perfectly into his dreams for the future.

Lincoln's father, James Ellsworth, had made his fortune in coal, and Lincoln was his only son and sole heir. Lincoln's mother had died when he was eight, and his last memory of her was that he took her hand when she was lying in bed, and she pulled it away. He barely saw his father in the years that followed, even when he wrote to him from boarding school and begged him to visit. "Dear Papa, please come and see me so the other boys will know I have a papa." His father never came. Many years later, Lincoln's widow found these letters, which she described as "really tragic letters," and burned them because she could not bear to keep them.[3]

When Ellsworth was a young man, his father tried to persuade him to go into the family business. He told him how much money he could earn and asked him to choose whichever home he wanted. His father invested in real estate and had a castle in Switzerland, two villas in Florence, and several properties in the US.[4] The houses were full of art and antiques and he was one of the first Americans to have a Rembrandt on his wall at home. His only heir replied: "I want nothing, but my tent and my gun."[5]

Lincoln invited his father and Amundsen for a meeting to discuss Amundsen's plan to fly to the North Pole. Having listened to Amundsen, the father asked the Norwegian what he would do if he did not give him the money, and the world-famous explorer replied: "What I have always done, I will get along some way."

For most polar explorers, it is more natural to do the opposite—first secure the financing, then decide to go. Ellsworth Sr. was impressed by Amundsen's attitude and promised to give $90,000, on the condition that his son stopped smoking a pipe, which Lincoln promised to do (but never actually did).[6] For Amundsen it was a given that it would be a Norwegian expedition, but he did agree that it would be called the Amundsen–Ellsworth expedition.

To show his gratitude, Amundsen gave Ellsworth Sr. his binoculars. Amundsen had used the Zeiss binoculars on all three of his expeditions, to the Northwest Passage, the South Pole, and Northeast Passage, and had the start date of each expedition engraved in the worn brass. Despite this generous gift, the paterfamilias regretted his decision soon after and tried to withdraw from the agreement. Amundsen was not a man to take unnecessary risks and so had gotten the agreement in writing before Ellsworth Sr. changed his mind. However, Senior did not give up and employed lawyers and a private detective in a bid to stop the expedition. The background for this was that he feared his son would die—which was perfectly possible—and that he himself would die from worry. He even sent his son a written warning: "If you persist it will cause my death."

The private detective and lawyers did all they could. They confiscated his passport and blocked a bank transfer of $10,000 from Ellsworth Jr.'s personal account. The father and his helpers reversed this when they were informed that part of the money was to be used to buy parachutes.

When Ellsworth Sr. realized that he would not be able to stop the expedition, he went for his son's jugular and said that the expedition was completely unsuited to scientific research of any value. Father and son were never reconciled after that. As Lincoln Ellsworth's friend Beekman H. Pool writes wryly in his biography of Ellsworth, *Polar Extremes*: "Amundsen accepted blame for splitting up a family, but concluded that circumstances made it necessary."

NOT ONLY DID Amundsen want to be the first to fly to the North Pole, but he did not want to turn back when they got there. Instead, he wanted to continue over the Arctic Ocean all the way to the gold-rush town of Nome in Alaska. Only when it was explained to him explicitly that it was in no way possible to carry enough fuel for such a long flight did he give in and agree to fly from Kings Bay, now Ny-Ålesund, on Svalbard to the North Pole and back, a distance of 761 miles each way. "Willpower and stamina could not overrule mathematics in the modern era," as Amundsen's biographer Alexander Wisting concluded.[7]

However, the polar explorer did not back down without expressing his disappointment in the new generation of explorers, and wrote in his diary: "I often ask myself what has become of the courage of youth." Deep down, he still hoped that he would be able to continue on over the Arctic Ocean, once he had reached the North Pole. He also dreamed of discovering an "oasis on the ice."[8] He was not alone in this. It is not unlike the vision of the North Pole in antiquity as a sunny region where life resembled the Garden of Eden. Furthermore, he had received instructions from the Norwegian government to claim sovereignty over any islands and land he might see from the air. Parts of the Arctic Ocean between the North Pole and Alaska were still unknown.

The flight was even more dramatic than Amundsen could have anticipated. On earlier expeditions, he knew all about the sledges, skis, clothing, food, dogs, and depots. This time, "the boss," as he was called, was the initiator, navigator, and fundraiser. The expedition was comprised of six men and two flying boats—*N24* and *N25*—which each had two Rolls-Royce engines. (A flying boat is basically a seaplane with a hull that can float on water.) Amundsen was the navigator of one plane and Ellsworth took charge of the other. Each had his own pilot. In addition, they both had a pilot and a mechanic on board with them.

At ten past five in the afternoon of May 21, 1925, the two flying boats—or whales, as they were sometimes called due to their appearance—took off from the ice by Svalbard. They flew north together; the men had worked out that it would take about eight hours to fly to the North Pole. To ensure that they would have enough fuel, they chose not to take heavy radio transmitters, which would have enabled them to communicate between themselves and with others.

A good eight hours later, they landed on a refrozen lead to establish their precise position. The ice is flattest on an open-water channel that has recently frozen over, and therefore best suited as a landing strip. Before *N25* had actually landed, one of the engines broke and the other was not powerful enough to keep the vessel in the air, so they had to make an emergency landing. *N25* managed to land without any significant damage, but *N24*'s hull had already suffered a tear on takeoff and the damage was too extensive to be repaired.

Both *N24* and *N25* were now on the ice, but any view of the way forward was obstructed by hummocks. The explorers could not see each other's planes, or the Norwegian flag that was then raised in both landing spots. When Amundsen climbed onto the wing of *N25* with his binoculars, he finally managed to spot the Norwegian flag by *N24*. The ice was moving all the time, and fortunately, the two flying boats were drifting toward each other. If they had been moving in opposite directions, it is not likely they would have seen each other again. Amundsen had written in his account of the South Pole expedition that luck and good preparation go hand in hand, but these favorable currents were purely good luck.

As the crew of *N24* trekked across the ice fields toward *N25*, two of them fell through the ice. Ellsworth managed to get down and lie flat on the ice before it gave way under him too. With the help of his skis, he managed to save the other two before they were pulled under the ice and lost for good.

Finally, five days later, all six men were gathered at *N25*, but their lives were in danger. They had not reached the North Pole, but found themselves at 87° 43′ N. As *N24* had been wrecked, they were dependent on *N25* being able to take off from the ice with all six of them on board. That seemed like an almost impossible task as the ice was uneven, flooded, and still moving.

On the ice-covered Arctic Ocean, Amundsen once again became the natural leader. His leadership style was very different from other polar explorers. Instead of the clear hierarchical structure of English expeditions, he chose a flat structure. He valued each member of the team and showed his appreciation when they did a good job. He also understood the importance of numbers, and that numbers could define a goal and inspire incredible motivation. And in this case, he linked the number to a date: June 15.

Amundsen chose that decisive date as the day they had to be ready to fly. If *N25* managed to take off from the ice with all six men on board before that day, they were likely to survive, according to the boss's calculations. If they didn't manage, all would be lost. They would need to cross on foot. They had skis and other equipment that they might need on the ice with them, but some of the men had very little experience of skiing, and they would have to cover around 620 miles on melting ice to reach safety.

They used the days running up to June 15 to make an airstrip, flattening the ice where it was uneven and laying old ice where there was too much surface water. They had wooden spades, knives, and ski poles. It was heavy and complicated work, as new hummocks were constantly forming where the men had leveled out the ice, and cracks appeared in other places. The daily food ration was reduced to 10.5 ounces per person. Amundsen estimated that they cleared around 500 tons of snow and ice. They also used their feet to trample down any loose, wet snow so the surface would be harder and allow the skis on the plane to slide more easily.

The day arrived, and conditions were good: The temperature was –27°F, so the ice was firmer, and there was a gentle southerly wind, which would help the aircraft to take off. The engines on *N25* were started and given forty-five minutes to warm up. The airstrip was 1,640 feet long and marked so that the pilot, Hjalmar Riiser-Larsen (1890–1965), could see how many feet were left, and at the right moment push the throttle. If he did not manage to take off in those 1,640 feet, they would crash into a hummock and presumably all die.

Riiser-Larsen had never been to the Arctic before, but had as much experience as a pilot from Norway could have. The pilot sat alone at the open front of the aircraft, with only a windscreen for protection. He had enough fuel to fly for eight hours. The five other men lay or sat as far back in the narrow plane as they could, presumably because Riiser-Larsen believed that the weight distribution would increase the likelihood of the plane being able to take off. They could not see anything. The men felt the airplane start to move forward with a jolt and then pick up speed down the airstrip until there was a final jolt and they realized they had taken off and were airborne. "The feeling of relief was indescribable," Amundsen wrote. He crawled forward to the cockpit and placed a piece of chocolate in Riiser-Larsen's lap as thanks. Then he thanked God for what seemed like a miracle.

Eight hours later they could see land, but the steering mechanism started to falter and Riiser-Larsen had to make an emergency landing once again, this time onto the open sea fifteen miles north of the island of Nordaustlandet in the Svalbard archipelago. The flying boat was rather unsteady on the water, but the sea was relatively calm on the final stretch past Sjuøyane into land. Nordaustlandet is a desolate place, with low mountains and long glaciers. It is hard to estimate how many days it would take to ski to Kings Bay, but certainly no fewer than ten. And someone like Ellsworth, who could barely ski, would struggle.

But there was no need to get out their skis. A whaling boat happened to be nearby when the flying boat reached land, and they offered to take the six men round the north end of Svalbard and south to Kings Bay. When the whaling boat passed Virgohamna, where Andrée, Frænkel, and Strindberg had taken off in *Örnen* in 1897, a flag was raised in their honor. Amundsen talks of his respect for Andrée in his book about the flight to the North Pole. Having attempted to fly to the North Pole himself, he says, there was no one on earth who had a better right to honor the man's memory than Amundsen and his five friends.

When the whaling boat docked in Kings Bay, people were so surprised that the members of the expedition were still alive that they had to touch the men's cheeks just to be sure. The expedition's storekeeper, the pharmacist Fritz Gottlieb Zapffe, Amundsen's friend and father of Peter Wessel Zapffe, was waiting for them. When he saw the polar explorer, looking exhausted in reindeer skin clothes and a sixpenny cap, he was shocked to realize that Amundsen had become an old man in those weeks on the ice. "The sight of him broke my heart," the pharmacist wrote.

The expedition had not made it to the North Pole and Amundsen must have been worried about how he would be received on his return to Oslo. To his astonishment, the expedition was even more widely celebrated than his previous expeditions, which had achieved their goals. It was an international sensation. When *N25* flew low up the Oslo Fjord and landed in the capital on July 5, 1926, the expedition team were welcomed by 50,000 jubilant Norwegians. It was apparently the largest gathering in the history of the capital, even more than for Nansen's return in 1896 and when the King of Norway, Haakon VII, arrived for the first time from Denmark in 1905.

The six polar explorers experienced what others before them had experienced: If you are away for longer than people at home think is possible to survive in a frozen wasteland, the shock and joy that you are still alive is almost as great as if you had been resurrected.

Lincoln's father was right, however, in one of his predictions. He died on June 2 while his son was battling to stay alive on the ice. Some see it as the father's revenge, but really it's a great tragedy that he did not live long enough for his son to tell him about his victory.

Long-Distance Love

In the months before the 1925 expedition, and in the weeks when they were caught on the ice, Amundsen's diary is a kind of continuous declaration of love to the English woman Kristine "Kiss" Bennett. He had been in love with her for thirteen years, but she was married, lived in London, and had two sons, whom Amundsen got to know well. On February 21, he wrote: "My life is you." In the middle of March, he expressed the need to be loved by the only person in the world "I really care about." Kiss Bennett believed she was able to communicate through the power of thought, and Amundsen writes: "Can you feel me? Kiss, Kiss, Kiss—this is our radio, can you hear me?" and carries on to tell her that he loves her with all his heart.

On the way to Kings Bay, before the start of the expedition, he writes: "Goodnight, my love. God bless you." When he does not hear back from her, he is fretful and at times jealous. The last thing he wrote to Kiss before *N25* took off on May 21, when he was clearly not sure that he would survive, was: "Adieu, my love."

Long-distance love worked well until Amundsen got the feeling that Bennett wanted to leave her husband in London and move in with him in Uranienborg. Then, his love cooled. Amundsen had a habit of falling in love with married women who lived far away from where he usually was, such as London or Alaska. Nansen also liked women who were already married. His sexual debut was with a priest's wife who was sixteen years older than him, when he was lodging in a rectory in Bergen. The relationship lasted for the three years he lived there, and he later had relationships with a number of married women.[9] Perhaps what made married women particularly attractive was that there was no need for commitment: It required no more than a few stolen hours and the ability to travel at short notice.

Amundsen has been criticized for loving Bennett so passionately before and during the expedition, but then falling out of love with her after. Perhaps those who criticized him did not understand the pressure he was under and how isolated he had been for years. The threat of bankruptcy constantly hung over him; he was ridiculed and belittled. When the idea of flying started to take shape, he knew there was a risk of death. On previous expeditions, his experience was the best guarantee of survival, but this time it was up to pilots who had never been in the Arctic, and airplanes that had never been used in a similar climate.

I have myself experienced how love can flare up when I am isolated, under pressure, without the possibility of correction from outside. When I walked to the South Pole, I had no contact with the world for fifty days, and I was in love with a woman I barely knew. She grew and changed in my thoughts from day to day, and the farther south I went, the more beautiful she became, and the more deeply I fell in love. It was comforting to have these thoughts and fantasies. But when I got back and we met again, reality shattered my illusion. It would have been impossible to live up to what I'd built in my mind.

The Corrupt Pole

The North Pole is, for me, the corrupt pole, or corrupt goal. Corrupt, not in the sense of bribery, but as a result of all the disputes that have followed in the wake of expeditions—the greatest, of course, being who got there first. And before that, there were accusations and disputes about the record for getting farthest north. To this day, explorers are accused of starting farther north of land, of not having gone to the North Pole itself, or of not having followed the ever-changing guidelines. Many have of course been able to document the truth of their claims, while others run into problems. In my experience, however, most people tell the truth, including polar explorers.

When we were starting our trek to the North Pole, the airplane landed on the water 980 feet from Ellesmere Island. To ensure that no one could say that the expedition had not gone all the way on foot, I put on my skis and went south until I was standing among bare stones on the shore. I put a stone in my pocket and then skied back to the others. We agreed that it was pretty pointless, but I could not bear the thought that someone might quibble that we had started too far north, after we had skied for 500 miles.

The pack ice is one reason why it is so hard to prove claims about having reached the North Pole. As it changes every day at the North Pole, it would be easy to disguise the truth. The ice is constantly drifting away from the polar point toward the Atlantic Ocean or the coasts that edge the ocean. Anything left at the polar point, such as a flag, an igloo, or a tent, is meaningless as evidence. Robert Peary wrote in his book that on April 6, 1909, he left a bottle between a hummock and a block of ice with a message in it to say that they had reached the North Pole: "I have with me 5 men, Matthew Henson, colored, Ootah, Egingwah, Seegloo, and Ookeah, Eskimos; 5 sledges and 38 dogs." Frederick Cook claimed to have done the same: "April 21 at the North Pole. Accompanied by the Eskimo boys Ah-we-lah and E-tuk-i-shuk I reached at noon to-day 90° N." They both knew that no one would find a message in a bottle at the Pole itself, but there was a slim chance someone might find the bottle on a shore farther south.

The South Pole remains the same, year in and year out. When I got there on January 7, 1990, the first thing I saw on the horizon was the Amundsen–Scott South Pole Station, a base built by the Americans at the polar point. Neither I nor anyone else could be in any doubt that I had gotten there.

Another reason why the truth about a North Pole expedition might remain hidden is that so many have died before they could tell their story. The numbers are uncertain, but in the period before the twentieth century, it is estimated that around a thousand men tried to reach the North Pole and sail through the Northwest Passage and the Northeast Passage, and of those, 751 died doing so.[10] It was a frighteningly lethal journey. The figure might be more accurate if it included those who died after giving up, such as Willem Barentsz and the members of Adolphus Greely's expedition.

You are more certain of reaching the South Pole, as you are effectively on a gigantic glacier. Expeditions take place in summer, and there are no dangerous animals or open water. The race to the South Pole only lasted a decade, as opposed to a century, and there have been far fewer expeditions in the south.

THE RACE TO be the first to fly to the North Pole continued in May 1926.

Shortly after midnight on May 9, the American naval officer and explorer Richard Byrd (1888–1957) took off from Kings Bay in *Josephine Ford*, a three-engine Fokker. The American pilot Floyd Bennett was at the controls and would steer the plane both ways, while Byrd navigated. The airplane was named after the three-year-old daughter of the main sponsor, Edsel Ford, the only son of Henry Ford. The *New York Times* had bought exclusive rights to the story. Almost sixteen hours later, they landed back at Kings Bay. Byrd could tell that they had reached the North Pole and circled around the top of the world for fourteen minutes. But he had forgotten to throw out the Stars and Stripes flag at the North Pole to mark that they had been there, and the flag came back with them.

The news was telegraphed to the *New York Times* from Kings Bay. And the front-page headline the following day ran over three lines: "Byrd flies to North Pole and back; Round trip from Kings Bay in 15 hrs. and 51 mnts.; circles top of the world several times."

The story quickly spread around the world to become the most repeated news since the end of the First World War. On their return to the USA, the city of New York put on a ticker-tape parade for the two men, in an open-top car up Broadway. They were also awarded medals at the White House by the president himself and given a promotion in the marines. Byrd spoke with a mixture of pride and exasperation about the whole thing as "the hero business."[11] A question, however, remained: Did Byrd fly all the way to the North Pole?

Seventy years later, I spoke to Helge Ingstad, the explorer who, together with his wife, Anne Stine, discovered Viking settlements in Newfoundland, the final proof that the Vikings had made it to America. Our conversation turned to the race to reach the North Pole first by airplane, and I suggested that Byrd was probably first. But Ingstad shook his head. He was born in 1899 and still remembered the expedition well. A friend of his, Bernt Balchen, who was a pilot, was in Kings Bay in May 1926, and had repaired the plane's skis after they were damaged on their first attempt to reach the North Pole. Byrd was so pleased with the Norwegian's work that he took him on as a pilot and mechanic, and later as the head pilot on his South Pole expedition. Balchen knew the Fokker planes better than anyone, and he and Bennett flew together in *Josephine Ford* for 100 hours and 55 minutes.

Unfortunately for Byrd, Balchen was not convinced he could have flown to the North Pole and back in such a short time, and over the years he gathered some good arguments. He racked up many more hours in the airplane than Byrd, and knew that the top speed was 74.8 miles per hour (65 knots) when the plane had landing wheels. They would have had to fly at 100.3 miles per hour (87.2 knots) to reach the North Pole and back within that time. In addition, the plane had skis rather than landing gear, which would have caused greater wind resistance and made it even harder to maintain that speed.

Balchen told Ingstad that Bennett had told him that they had been nowhere near the North Pole, but had flown back and forth north of Svalbard, then turned and headed south.

One detail that speaks in Byrd's favor is that Amundsen, who also was in Kings Bay, was quick to congratulate and kiss him on both cheeks, as soon as they met. However, in private, he did express his doubts that Byrd managed to reach the North Pole.[12] Byrd's supporters have argued that there was a strong tailwind blowing on the way north and again on the way south, which accounted for the speed, which is not impossible, but does seem unlikely.

THE DREAM OF being the first to cross the Arctic Ocean, and perhaps even discover new territories, lived on in Roald Amundsen and Lincoln Ellsworth. They were both in Kings Bay when Byrd was getting ready to fly north. Without having said that he no longer believed Cook's claim to have reached the North Pole, Amundsen had come to the conclusion that Peary must have been there. For him, the most important race now was to be the first to cross the Arctic Ocean, via the North Pole.

Having barely survived the *N25* venture, it seems the Norwegian wanted to take fewer chances with unstable engines, mist, and insufficient fuel. His chosen mode of transport this time was another technological innovation: the airship. As so often, he lacked the finances. Ellsworth was happy to contribute a substantial amount, but not all of it. Ellsworth was one of the few explorers in the history of the North Pole who did not, like Børge and me, beg sponsors for money.

Two companies in Europe were spearheading the development of the airship. One was Luftschiffbau Zeppelin in Germany, the leading manufacturer at the time, and the other the Italian State Airship Factory. Amundsen's reputation in Germany had suffered somewhat after he returned his German honors in 1917 "as a personal protest against the German murders of innocent Norwegian sailors."[13] Fridtjof Nansen did not return his honors. One consequence for Amundsen was that it complicated working with the Germans and being able to use one of their well-constructed zeppelins.

The alternative was Italy. Umberto Nobile (1885–1978) had visited Amundsen in Uranienborg in July 1925 and offered to design and build an airship for him. The engineer was a world leader in airship design and Amundsen accepted the offer. The Italian prime minister, Benito Mussolini, and his fascist government gave a guarantee that they would buy the airship back after the expedition, which was an important stipulation for the funding. As Amundsen had said to Ellsworth Sr., when it came to money: "I will get along some way."

The cigar-shaped airship was christened *Norge*; Amundsen wrote that he naturally had "stipulated that the ship not only would fly the Norwegian flag, but also . . . carry my fatherland's name." The Amundsen–Ellsworth–Nobile Transpolar Flight took off from Kings Bay at 9:55 a.m. on May 11, 1926. As they traveled, Amundsen sat on a velvet-covered chair made from the newly invented duralumin and kept a lookout for land in the Arctic Ocean. The fate of the expedition was more in the hands of the pilot, Riiser-Larsen, and engineer, Nobile, than in his.

The relationship between the Italians and the Norwegians was terrible. The expedition was viewed very differently by the two countries, even if the goal was the same. In Italy, it was presented as a fascist sensation, and Nobile was hailed as the great innovator and leader. In Norway, it was a Norwegian expedition, led by Amundsen, the polar hero.

When *Norge* passed 87° 30′ N at ten o'clock in the evening, Amundsen received a private telegram. Being at the ice cap no longer entailed isolation. The telegram contained the news that his two champions in Buenos Aires and Chicago had finally bought his home in Uranienborg. Amundsen wrote briefly in his diary that the buildings on the property were now his.

The crew of the Norge *hauling the giant airship out of the hangar for her start to the North Pole.*

Lincoln Ellsworth in snowshoes and Capt. Roald Amundsen on skis, the American and Norwegian leaders of the flight of the Norge *over the North Pole, on Spitsbergen Island in 1926.*

A good four hours later, on the night of May 12, *Norge* reached the North Pole, and stopped 556 feet above the ice. Amundsen, Ellsworth, and Nobile got ready to drop their country's flags down onto the Pole, in the agreed order. First Amundsen dropped a Norwegian flag made from silk, then Ellsworth followed suit with the Stars and Stripes. Amundsen writes in his memoir, *My Life as an Explorer*, that Nobile had insisted that the flags should be small, in order to minimize weight, and they had both honored this request. The flags were about the size of a handkerchief. Then it was Nobile's turn. And Amundsen writes that suddenly *Norge* turned into a celestial circus, with flags in all manner of shapes and colors floating in the air around them. He describes the Italian flag as colossal.

Col. Umberto Nobile, designer of the dirigible Norge, *walking along the catwalk inside the huge gas bag during the flight over the polar area, on July 14, 1926.*

Despite weight allowances, Nobile had his fox terrier, Titina, with him. He estimated that she weighed around eleven pounds. Titina had jumped into the gondola without him noticing, he said, but none of the Norwegians believed him. The fox terrier was wearing a woolen coat and was left to run about between their legs. Amundsen joked with a wry smile that it might be handy to have dog cutlets in case of an emergency, a dish he had simply called "dog" fifteen years earlier when he went to the South Pole.

When the airship reached the North Pole, there were in fact two men on board who had then been to both poles. The other was Amundsen's friend Oscar Wisting, who had also accompanied him to the South Pole. Amundsen offered him his hand. To share the honor of being the first person to reach both poles, when you do not need to, shows a magnanimity that is rare in explorers.

Ellsworth, for his part, felt not only joy at reaching the North Pole, but also disappointment. The American president had given him a Stars and Stripes to drop down, as the first American to fly the full distance, but now Richard Byrd was officially the first.

It is dispiriting to think that everyone on board died before they could be honored in the way they deserved. If Cook or Peary had not actually reached the Pole, and Byrd had not actually flown over it, then the Italians, Norwegians, and the American were the first in the world to stand directly above the North Pole.

An unknown oasis was never discovered, but the explorers showed that it was highly unlikely that there was any land in the Arctic Ocean. *Norge* flew into a thick mist between the North Pole and Alaska, and the balloon was covered in frost, which then turned to ice that started to fall off into the propellers, and then hurtled down toward the men in the control car. For the second time in less than a year, things could have gone seriously wrong for Amundsen and Ellsworth.

The radio on board stopped working, and a considerable share of the world held their breath. On board, Amundsen seemed to be happy about it, according to Nobile, who suspected that the Norwegian had sabotaged the radio so he could, once again, surprise everyone by having survived.

Norge got to Alaska, but not to Nome. Instead, the airship landed after seventy-two hours in the air in a village called Teller. No sooner had they landed than the Norwegians and Italians started to compete in trying to get their story out to the world first—and then to be believed. Nobile felt that he had, in fact, been the real leader, while Amundsen claimed that Nobile had on several occasions been so tired that he almost fell asleep, and that *Norge* had been in danger of crashing as a result, and that he was the actual leader. Nobile and Riiser-Larsen apparently got so angry with each other at one point in the air that they started to fight in the gondola.

It is not easy to know who was telling the truth, but I think neither of them was right. Nobile had designed and built the airship and flown it from Rome to Kings Bay, without Amundsen. He knew the aircraft well. Amundsen, on the other hand, felt he had spent his whole life and all his money and time on expeditions, so without him they would never have reached the North Pole in 1926. Whatever the case, back in Norway there was never any doubt that it was the Norwegians who should be celebrated.

In Italy, with Mussolini at the helm, it was Nobile's genius that was celebrated, his leadership and courage. The dictator was more gracious, however, when he spoke to the right-leaning Norwegian newspaper *Aftenposten*, crediting "the unconquerable courage of the Italians with the iron will of the Norwegians, and the virility of the Latin man with disciplined artistry of the Nordic man."[14]

Nobile was made a general after the expedition and Mussolini said in the ceremony that it was Nobile who deserves all honor. "You, an Italian, designed the airship. You, an Italian, brought other Italians together and steered them to the goal of this magnificent voyage."[15]

The Norwegians' scorn for Nobile was so deep and lasted for so long that his book, giving his version of events, was not published in Norway until 1976, fifty years after the expedition.

ROALD AMUNDSEN WAS loyal to those who stood by him, but often experienced the opposite himself. Having made himself unpopular in Germany, he became even more unpopular in Italy. Perhaps he felt that less fame and therefore less income was an acceptable price to pay for being able to speak his mind.

Roald Amundsen's book *My Life as an Explorer* was published the year after the expedition and caused a fall in popularity for Amundsen in the US and Britain as well. Amundsen criticizes American journalists, questioning their credibility, but reserves his fiercest attack for Europe's superpower. Amundsen explains that he has the utmost respect for Robert Falcon Scott as a "splendid sportsman as well as a great explorer." No one has admired British explorers as much as he does, he says, but he has no respect whatsoever for the men who stayed at home or sat in the offices of the Royal Geographical Society and failed to understand what happened on the way to the South Pole, or along the Northwest Passage. His conclusion is unambiguous: "By and large the British are a race of very bad losers."[16]

Amundsen was, by then, arguably the greatest living explorer and famous throughout the world. Norway's ambassador to Great Britain wrote to Nansen that it was not "easy to measure the damage that Amundsen has done" in terms of wrecking Norway's worldwide reputation.

Nansen replied that Amundsen must have "mental issues" and said the same to his friends in the Royal Geographical Society.

The gentlemen of the Royal Geographical Society had been

disparaging about Amundsen's South Pole expedition because they had used dogs to pull the sledges. Horses and motorized support were acceptable, but the age-old skepticism of dogs lived on. It was, of course, an embarrassing defeat for the British that the Norwegians had gotten to the South Pole first. However, the society did invite Amundsen to give a lecture in its august premises in west London. It was and still is a tradition and an honor for explorers. When the explorer finished his lecture, the president of the society elected not to say anything positive about the Norwegians' achievement, though he had for Nansen and Peary, and likewise for Børge and myself eighty years later. Instead, he chose to mock the Norwegians, as Amundsen remembers it; he raised his glass and said sarcastically to those assembled: "I therefore propose *three cheers*, for the *dogs*."[17]

Roald Amundsen's Final Journey

In Norway, the contempt for Nobile refused to dissipate. In fact, two years later, in 1928, *contempt* might be too weak a word. Many Norwegians had started to hate Nobile.

The Italian was back in Svalbard with the airship *Italia*, which he had designed and built himself with financial support from his motherland. The aim once again was to reach the North Pole, and there was little doubt that he and Mussolini's government would share the honor.

Italia reached the North Pole on May 24, and Nobile threw out the Italian flag and a cross that Pope Pius XI had asked him to leave at the Pole. The intention was to be the first expedition to land at the North Pole, and to erect the cross on the ice. But conditions were such that the airship could not land. On the return flight to Kings Bay there were strong winds and poor visibility, and *Italia* crashed 62 miles north of Nordaustlandet. The gondola hit the ice, killing one man on impact, and the balloon was torn loose from the gondola and sailed on with six men, who were never seen again. The radio was not broken, so the Italians were able to let people know they had crashed and roughly where they were.

On May 26, Amundsen was contacted in Oslo by the Norwegian minister of defense, Torgeir Anderssen-Rysst, and invited to the minister's office that same day. Amundsen's immediate reply—"Right away"—was in English, as though he intuited international interest. The minister informed him that Norway had received an official request from Italy to organize a rescue operation for Nobile, and he asked Amundsen if he could lead the operation, and if so, when he could leave. It must have been glorious for Amundsen to be given the opportunity to humiliate his famous rival, by saving his life.

The pleasure was short-lived, however. There was a message from Italy a few days later to say that Mussolini no longer wanted a Norwegian rescue operation. National interests were always paramount for Mussolini, and they might be better served if Nobile became a martyr in the Arctic, rather than returning home a failure.

But the eyes of the world were once again on the missing polar explorers. Around a hundred rescue operations were organized in six countries that summer. And around 700 people took part in those operations. Large ships, smaller vessels, dog teams, skiers, and airplanes were involved. It became a race between nations and men, not dissimilar to the race to find John Franklin's expedition. In private, Nansen commented that "the world is getting a little hysterical." He was right: Nobile couldn't be suffering too much, as they had enough food with them.

Amundsen was loath to give up the dream of saving Nobile, even though the Italian authorities no longer wanted him, and Nobile most certainly didn't. Nor did he let the fact that he was now engaged to another long-distance love, who was on her way from Alaska to Oslo, stop him. His fiancée was called Elizabeth "Bess" Magids and has been variously described as attractive, an astute businesswoman, and a mean poker player. She had been married when they met, but was now divorced and ready to move in with Amundsen.

He had been involved with Magids before his affair with Kiss. When the relationship between Amundsen and Kristine "Kiss" Bennett was at its peak, his relationship with Bess cooled, only to flare up again in the autumn of 1927. Bess stayed with Amundsen in Uranienborg, and he looked after her impeccably. In my experience, when you have lived at such close quarters with others as Amundsen had on all his expeditions, you become good at taking care of people—for as long as the journey lasts.

As ever, Amundsen lacked the money and, once again, a solution appeared. A Norwegian wholesaler in Paris with good contacts in the French government managed in the course of a few days to persuade the French Navy to grant Amundsen use of a Latham 47 flying boat, with a French crew of four.[18] The flying boat was white, with double wings and a wingspan of eighty-two feet. Amundsen was informed of the offer on June 14, and two days later, on June 16, a Latham flying boat flew from Normandy to Bergen, where Amundsen was waiting to board.

Amundsen invited a Norwegian pilot, Leif Dietrichson, to join them. Unlike the Frenchmen, Dietrichson had experience. In 1925, he had been the pilot of *N24* and was one of the two men whom Ellsworth saved when they were about to vanish under the ice.

The Latham 47 landed in Tromsø harbor at six o'clock in the morning of June 18. Bess had already left Alaska and was making her way east, by train and boat, to Norway. The Norwegians had breakfast at the home of the pharmacist Fritz Zapffe—a short rest before the expedition north. It was raining and the forecast was for deepening low pressure farther north. The flying boat was carrying 396 gallons of fuel and lay heavy in the water—with hindsight, they should have postponed their departure. But the desire to find Nobile before anyone else was pressing.

At four in the afternoon, the Latham 47 took off and headed north. The last sign that the six were alive was at 6:45 p.m., when a telegraphist in Tromsø heard a message being sent from the Latham to the radio station in Kings Bay.

No one knows what happened after that.

From the first days of silence, the Norwegians refused to believe that Amundsen might be dead and continued to wait for his return. The hero had always come home before, after all. Two weeks after Amundsen had disappeared, Bess arrived in Oslo and moved into his home in Uranienborg. After a week by herself in his house she moved to a hotel. Five weeks after her fiancé's departure, she gave up and went back to Alaska. Helge Ingstad told me that he had spoken to her daughter, who had said that her mother had been proud and glad of the time she spent with Amundsen.

In the days and weeks that followed, the Norwegians continued to hope. But then, on August 31, a boat fishing for halibut off Bjørnøya found the flying boat's left float. A 7.9 × 7.9 inch hole in the metal indicated that the flying boat had been damaged on landing. But it had not been a crash landing. An attempt had been made to repair a strut that ran through the float and the aircraft. On October 13, one of the fuel tanks was found by a fishing vessel by Haltenbanken, off Trøndelag. There was a wooden plug in the filling nozzle, a sign that the crew had tried to stop the tank from leaking. The repair may have been an attempt to replace the float that had been damaged on landing.

The Latham's floats were known to be a weakness in the flying boat's design. The French second pilot, Albert de Cuverville, was experienced in finding temporary solutions. On an earlier occasion in the Mediterranean, he had removed the fuel tank, plugged it, and attached it as a float, all while on the water. They had then managed to get the flying boat back into the air.

When the Latham presumably landed on the water on June 18, a life-and-death struggle would have ensued that may have lasted hours or even days. It is likely that de Cuverville detached the damaged float and replaced it with the fuel tank, while the flying boat rocked on the ocean. I have studied the modified fuel tank, which is in the Maritime Museum in Oslo. A shiver ran down my spine when I saw three square holes in the tank, presumably cut by de Cuverville so he could attach the tank to the bottom of the aircraft. If he had succeeded in doing this, the pilot must have tried to start the engines again and allowed them to warm up before accelerating across the surface of the water and pulling back the gear sticks when he believed they were ready to take off, just as Amundsen and Dietrichson had experienced in 1924 when *N25* took off from the frozen Arctic Ocean, but this time without the good luck they had relied on.

ONE NOVEMBER EVENING in 2022, I went for a walk from the Fram Museum along the shore at Frognerkilen with Tor Bomann-Larsen, one of Roald Amundsen's biographers. Bomann-Larsen commented that in later life Amundsen had become more and more like his role model, John Franklin, as well as like Salomon August Andrée—two heroes who he believed had died as martyrs. As a young man, Amundsen's dream was to fulfill Franklin's goal of sailing through the Northwest Passage. And later in life, he would realize Andrée's dream to fly to the North Pole and farther across the Arctic Ocean. He survived as long as he did thanks to careful preparation, but through the 1920s he started to take less care. It was as though he had just grabbed the chance to go north in the Latham. A bit like Andrée in 1897. "In a way, all polar explorers have sought the kingdom of death," Bomann-Larsen said, "in the sense they tried to outsmart the elements. But I think it's true to say that Amundsen, with his flights, almost challenged death in a calculated manner, or to put it another way, put his life in God's hands."

The mystery surrounding the Northwest Passage, the South Pole, and the North Pole had fueled Amundsen's fantasies. As with other explorers, the line between knowledge and fantasy was blurred before he'd even set out on an expedition. It is often during the preparation process that knowledge took over from the dreams. But at the end of his life it seems that the dream of rescuing Nobile from the ice was more important than any hard-won knowledge of flying, ice conditions, and weather forecasts.

Amundsen believed he would survive once again, although there are a number of things that might indicate otherwise. Even before Nobile flew north in *Italia* and before Amundsen could dream of any rescue operation, he was interviewed by an Italian journalist for the *Corriere della Sera* newspaper. He talked about the Arctic Ocean in the interview and told the journalist: "If only you knew how magnificent it is up there! That is where I want to die." Amundsen went on to say that he hoped he would die "in an honorable manner" while carrying out a noble task. And ideally he would die quickly, and without suffering.[19] Both Zapffe and fellow explorer Wisting said that Amundsen was resigned even before he left.

At the beginning of 2000, I asked at the Polar Museum in Tromsø if they knew of anyone in the city who had seen Amundsen on June 18, 1928. The director of the museum, Torbjørn Trulsen, told me there were two people who were still alive. I talked to one on the phone and then traveled to Tromsø to meet the other. Together we walked the same route through Tromsø that Amundsen had taken from Zapffe's apothecary down to the water and along the harbor.

He described how Amundsen had walked with a light step and waved happily to the locals as he passed. He waved his hat to the crowds from the Latham, acknowledging his popularity. Both witnesses recalled that Amundsen looked like he was glad to be going on an expedition again. Whatever the case, it's difficult to believe that Amundsen would take the Frenchmen and his friend Dietrichson if he knew he was going to die. As a leader, Amundsen placed great importance on the safety of his men.

ON OCTOBER 25, 1928, when all hope had been abandoned, Fridtjof Nansen held a memorial speech for Roald Amundsen. Nansen is effusive in his praise: Amundsen was now a dead hero and so easier to relate to. Nansen talks about courage and refers to the historian Thomas Carlyle and his statement that "the first duty of man is to conquer fear." Amundsen had felt that fear, just as Nansen himself had and almost anyone who has prepared for a North Pole expedition has.

There is something brilliant about the way in which Amundsen prepared himself for expeditions, and everything that Nansen achieved, but neither of them found it easy. Their lives were haunted by a perpetual fear of failure, of losing respect, combined with a constant need to break barriers and strive for something greater and beyond themselves. Neither of them was ever able to rest.

Nansen must have realized, in giving his speech, that Amundsen had finally won the competition to be the greatest polar explorer ever. Nansen was by then an old man and would die in his own bed. Amundsen had disappeared in the icy wastes, a martyr, while on a final noble expedition.

Many Norwegians were in no doubt about who was to blame for Amundsen's disappearance. It was Umberto Nobile. For them, if he had only stayed away from the North Pole, the tragedy would never have happened.

In the summer of 1928, Norwegian newspapers were full of stories about rescue operations. First for Nobile, then for Nobile and Amundsen, peaking when Nobile was finally rescued after forty days. He was found on the ice with eight other survivors, but the airplane that found them only had room for one passenger. As expedition leader, he was rescued before the others, and took his fox terrier, Titina, with him.

Amundsen's opinion of Nobile is summarized in the three-volume work *Norwegian Polar History*: "Vain, arrogant and conceited, a show-off and foolish dreamer, uniform-loving, nervous, ridiculous and spoilt, and a hopeless navigator." Most Norwegians shared this opinion. In fact, the hate was so widespread that the rescue ship, *Città di Milano*, which took the survivors back to the mainland, did not dare take them to Tromsø, which was the closest port. Instead, they sailed to Narvik, which had a direct train to Sweden.

Nobile was portrayed as the Judas of polar exploration in Narvik's local newspaper, *Fremover.* But according to the journalist Nobile was even baser than Judas, because the disciple at least had enough self-respect to hang himself after his betrayal, whereas the fascist was sailing toward the Fagernes quay in Narvik harbor. He was a "pathetic mummy's boy" and a pompous general with "a cardboard heart and cork veins."[20]

Città di Milano docked early in the morning of July 26, and in the afternoon a rail carriage was driven right up to the boat. The gangplank was run from the boat into the carriage. Norwegians did not want any of the Italians to put foot on Norwegian soil, and thanks to the gangplank from the boat to the train, none of the nine were given the opportunity.

When Nobile got back to Italy, he was considered a failure and had fallen from favor with Mussolini. It was hard to continue living there when the dictator was constantly disparaging him. He first moved to the Soviet Union and from there to the US. He worked on the national airship program in both countries. The Russian airship he helped to build crashed in 1938, but there were no accidents in the States. He returned to Italy in 1946 and was rehabilitated. The same year, he stood as the Communist Party's candidate in the general election and was voted in.

ANYONE INTERESTED IN Norwegian polar history tends to say that Amundsen disappeared in 1928, rather than that he died. Dreams of his survival continued to circulate. As late as the Second World War, when Norway was invaded by Hitler's troops on April 9, 1940, there was a rumor that Amundsen was on his way back to the north of Norway with an Inuit army to liberate his country. And in 2017, the Norwegian polar explorer Monica Kristensen published a book to great critical acclaim in which she tries to prove that Amundsen survived and built a base on Bjørnøya. One of Norway's most popular clothing brands is named after Amundsen. The country will not let go of him.

Peter Wessel Zapffe—Polar Philosopher

Peter Wessel Zapffe was that rare thing: a philosopher with an interest in polar exploration.

In Zapffe's own words, he became a philosopher through panic rather than choice. And much of the panic was due to his father, who was a great admirer of Amundsen. He was hard on his son, who was unathletic and extremely shortsighted. Peter Wessel Zapffe was whipped and punished and subjected to what today would be called gross abuse.

His father went to Svalbard with Amundsen in 1925 and 1926 as a storekeeper, to support preparations for the flights to the North Pole. The polar explorer was a close friend and stayed with the Zapffes whenever he was in Tromsø. On the last morning that Amundsen was definitely alive, he had breakfasted with the Zapffes, as mentioned, before carrying on north and disappearing into the icy wastes. The son wanted to go north in 1928 to search for Amundsen, but was not able to join any of the rescue missions, and in 1930 he went to Kvitøya to inspect Andrée's last base. Peter Wessel Zapffe became a great outdoors man, but nothing he did satisfied his father. Fifty years after Amundsen reached the South Pole, Zapffe defined the criteria for being a great polar explorer, unlike himself, and more like Amundsen: "Three factors combine in this historical person: the man, the mission and the means. Each of them is a necessity, none is a sufficient prerequisite."[21]

When he was in his early teens, Zapffe asked Amundsen how he felt when he was standing at the South Pole. Nobody else had asked the question. Amundsen looked at him for a long time, without giving an answer. His expression and the lack of response made Zapffe think he probably felt nothing at all.[22]

Perhaps Zapffe was right: Amundsen did not spend time exploring his feelings. When the expedition reached the Pole, they spent their days documenting that they were there first, so that no one who came after them could claim otherwise. Amundsen wrote a message to Captain Scott and a letter to King Haakon that he then left in the tent that remained standing at the South Pole. Amundsen asked Scott to take the letter to the king with him when he went north again, in case he didn't make it home.

In *On the Tragic*, his doctoral thesis, Zapffe writes about the importance of not taking shortcuts when striving to reach a goal. When going to the North Pole, one must have as little help as possible, as the experience of standing on the top of the world will be superficial if the way there has been too easy. I adore that thought. People's need to make everything easier is destroying our opportunities for great experiences. Zapffe's basic premise here is that to achieve with effort is equal to the value of achieving something. One must struggle.

In his thesis, Zapffe explains why the life of a hero will always end in tragedy: Every time he or she achieves something great, people will want more, and sooner or later the hero will not want or be able to deliver. Effort and records are soon forgotten, even when they are recorded in the history books, as happened to Roald Amundsen around 1922 when Norwegians forgot about his expeditions and talked only about his impending bankruptcy. Or the hero tries to do something even more challenging and fails. Or starts to cheat and brag to make the feat seem even more spectacular, as Peary, Cook, and Byrd most likely did when they each claimed to have been at the North Pole.

Zapffe's main theory was that we humans are overequipped for what society expects of us. Life entails a "tension between tasks and ability" and our abilities are overdeveloped in relation to the lives we lead. We search for meanings in life, but fail to find them and waste our abilities talking about trivialities. He believed we humans therefore are not suited to the civilized world. In order to understand oneself and the cosmic prerequisites for life, it is necessary to distance oneself from the cities, to be in the wilderness, to face danger, and to live through great existential challenges.

According to the philosopher and farmer Sigmund Kvaløy Setreng, Zapffe talked about his arrival at the inner pole toward the end of his life—a pole that he believed was even more challenging to reach than the North or South Pole. As with the other two poles, once you are there, every step you take forward is a step back and every step takes you to a position where you have already been. Thus you can never return. "So one *stays* at the inner pole." I believe Zapffe thought that everyone should choose their own path and try to discover their own pole—a personal North Pole that does not exist on the surface of the globe, but only in your own mind.

Zapffe's feeling of reaching an inner pole is not unique to him. The Sufi Suhrawardi envisaged a similar pole, and the philosopher Henry Corbin wrote about it as well. Kvaløy Setreng, who was a close friend of Zapffe, maintained that the existentialists in continental Europe were too cut off from nature to truly understand Zapffe's reflections on the inner pole.[23]

In a later edition of *On the Tragic*, Zapffe added that if, contrary to expectation, you did come to understand the mystery of life, you should, despite your physical powerlessness, "grab the mystery by the neck and shake it like a dishcloth."[24]

The First to Reach the North Pole

Throughout the nineteenth century, the North Pole was the greatest of all geographical goals: greatest in terms of the resources used to get there, greatest in the number of lives lost in the struggle to get there, and greatest in terms of international media interest and books published. Not one of all the people who dreamed of getting there succeeded. It was to the nineteenth century what landing on the moon was to the twentieth, and what populating another planet is to the twenty-first.

In 1948, there was a top secret expedition to the North Pole. Like other top-ranking military officers in the US and Britain, the dictator Joseph Stalin believed that the Arctic Ocean would be of crucial importance in the event of a Third World War. The logic was simple: the superpower that wanted to attack the other would send airplanes and bombs via the shortest route.

Already in the summer of 1945, the British sent Lancaster bombers north to gather meteorological data and do reconnaissance for any islands where the military could build secret bases. The bombs that were normally carried on board were replaced by fuel so the airplanes could circle the Arctic Ocean for as many hours as possible. The USA used B-29 Superfortresses for their secret operations in the Arctic region in 1946, the same type of airplane used to drop the atomic bombs on Hiroshima and Nagasaki.

On April 23, 1948, at 4:44 p.m. Moscow time, three Soviet planes landed at the North Pole. The expedition was led by Alexander Kuznetsov and was made up of twenty men.[25] The airplanes were Li -2s, copies of the American DC-3. The Russians were possibly the first in history to stand at the North Pole, without them realizing it. Even though there were many who doubted that Robert Peary had reached the North Pole, there were many more who believed the American. It was only some decades later that people began to question his claim to have been the first. Whatever the case, the red Soviet flag, with its hammer and sickle, was planted in the ice and the expedition team celebrated being the first people from the Soviet Union to stand on top of the world.

The purpose of the expedition was to measure temperatures and magnetism, to study the weather, test the water, discover any unknown islands, and establish how deep the ocean was. They measured a depth of 13,251 feet, which was officially increased in 1956 to 13,408 feet, and in 2007 to 13,980 feet, so they were close. They set up base on old ice that was sixteen feet thick, which in my experience is safe. Even when there is violent movement in the pack ice, it should not break up into floes, but in 1948 the ice was more unstable than usual. Eleven years earlier, on May 5, 1937, one of the pilots had landed twelve miles from the North Pole and he compared the conditions. In 1937, the Arctic Ocean had been covered by solid ice, but now there was a lot of open water. As he said, at the North Pole, the Arctic Ocean "looks like a flood."[26]

After all the men had gone to sleep in their sleeping bags, with the exception of the one left on watch, the ice beneath their tents started to move. The watchman raised the alarm, but before everyone had managed to get out of their sleeping bags, the airplane was already sliding down into a new lead. One of the plane's two skis was still on the ice, and the other was sinking fast in the freezing, deep water. The pilot managed to start and warm up the engine and then fly one hundred feet away from the open water. During the night, the ice moved so much and broke up so quickly that it resembled "a store window struck by a rock," as a journalist from *Pravda* described it. The journalist went on to say that no matter what fate awaited the men, the Soviet flag would fly over the North Pole.[27] It took eight hours of toil and struggle for all the men to move the base and the airplanes to safe ice.

But no sooner had the exhausted pilots and scientists got back into their sleeping bags when the alarm was raised again. The ice moved so fast that it "was like being on a river when the ice breaks," and so it continued through the morning. The sixteen-foot-thick ice broke up completely, the base was divided in two, and the two airplanes were once again in danger of disappearing into the deep, black water. All the food and equipment was taken out of the planes and stored on the ice. The Soviet flag that had been raised on the first day was now flying on another ice floe that drifted onto the horizon, then disappeared.

Over the next twenty-four hours, the ice floe, with the men and the airplanes on it, drifted nine miles south from the North Pole. On the morning of the third day, something like a miracle occurred. When the height of the sun was measured with a sextant, they were once again at the North Pole. The southerly current had turned and their ice floe had now drifted as far north as they had drifted south.

When the Russians then looked to the horizon that same morning, they saw the Soviet flag drifting back toward them, until finally the two ice floes bumped into each other again. That was how the Russians described the movement of the currents, but when you are on an ice floe, surrounded by other ice floes, it is not always easy to know if it is the one you are standing on that is moving, or the others around you, or all the ice—like a parallel train departing the platform. And whether they are drifting north, south, west, or east. The Russians might well have been drifting toward the flag.

The story of the twenty Russians remained secret as long as Stalin was alive. The expedition crew were not allowed to talk about what they had experienced at the North Pole to anyone. It was not until 1956, three years after Stalin's death and eight years after the expedition, that the *Pravda* journalist could finally write about the three days at the North Pole.

On February 18, 1971, the *New York Times* published a press release from Reuters, which had in turn received a press release from the Russian news agency TASS, saying that the pilot Ivan Cherevichny had died at the age of sixty-one. The short article contained information about Cherevichny, who "flew to remote Arctic stations and took part in rescue operations in the extreme north" and had been honored as a hero of the Soviet Union. The newspaper did not mention that he had been captain of the first of the three airplanes to land at the North Pole.[28] I have read countless articles about who was first to the North Pole—Cook, Peary, Byrd, or Amundsen?—and the leader of the Russian expedition, Alexander Kuznetsov, and the pilot, Ivan Cherevichny, are rarely if ever mentioned.[29]

Every compass bearing, unless it is straight east or straight west, will eventually lead you to the North Pole or the South Pole. If you walk, sail, or fly the course at 10°, 30°, or 60° N, you will circle the polar point until you get there, on the globe or in reality. The point that any bearing north leads to was the point that probably no one had reached before, until the three Russian airplanes landed there on April 23, 1948.

After all the dramatic, life-and-death accounts of men's attempts to reach the North Pole, it seems strangely fitting that no one paid attention when an expedition actually stood on top of the world. With the world's gaze elsewhere, these Russians quietly, finally, stepped on the most northerly part of our world.

THIS LACK OF certainty in connection with expeditions to the North Pole is not exclusive to the nineteenth and early twentieth centuries. On August 2, 2007, the Russian Arktika 2007 expedition captured headlines around the world. For the first time, people had reached the ocean floor under the North Pole, a kind of fifth North Pole. Or as they claimed: The Russians were the first to reach the *real* North Pole. That is to say, the point where the rotational axis from the celestial North Pole hits the seabed and pierces the surface of the earth, not just the Arctic Ocean, and then continues on down to the South Pole.

A Russian flag, made of titanium to withstand the pressure and ocean currents, was placed on the seabed and the message to polar explorers around the world was clear: "If a hundred or a thousand years from now someone goes down to where we were, they will see the Russian flag." The three Russians—Anatoly Sagalevich, who was both the captain and the pilot, the businessman Yevgeny Chernyaev, and the polar explorer Artur Chilingarov—used a very special submergence vehicle, the *Mir-1*. The submarine has a maximum operating depth of 1,970 feet and was built by Sagalevich. Half an hour after *Mir-1* touched down on the seabed, her sistership, *Mir-2*, followed, with the pilot, Genya Cherniaev, the Swedish explorer Frederik Paulsen, and the Australian explorer Mike McDowell on board. The six combined had considerable experience of the North Pole and deep dives with small submarines.

The Swede and the Australian were surprised to discover that the news about their expedition included a geopolitical dimension. Finally, it might be possible to prove that the Lomonosov Ridge, an underwater mountain chain that stretches 1,120 miles along the ocean floor from east to west, under the North Pole, was part of the Russian continental shelf.

There were loud protests from the US, Canadian, Norwegian, and Danish governments. Denmark and Canada believe that their continental shelf stretches from Greenland and Canada to the North Pole and do not accept any other nation's claim. The Canadian foreign minister, Peter MacKay, included a line from the country's national anthem in his counter-argument: "This is the true north strong and free!" before stating that marking the discovery of new territories with a national flag was a practice that belonged to the Middle Ages. The Russian foreign minister, Sergey Lavrov, who praised the Russians' courage, gave an amusing reply: Explorers have always raised their country's flag when a geographical goal has been reached for the first time, and "such was the case on the moon, by the way."

MacKay and his colleagues could, had they known better, have explained that Arktika 2007 was an international expedition, and not a Russian one. Plus it was a private expedition, not a state one. McDowell had initiated the idea and the Swede, Paulsen, had covered 90 percent of the costs from his own pocket. As Paulsen said when we spoke, he had two advantages that no one else had: money and the right contacts.[30] He explained that the Russians, Mike, and himself had not considered that the expedition might be of political significance until they were on board the ship that transported the submarines from Murmansk to the North Pole. Glasses were raised as they traveled north, and the idea cropped up after a few bottles of vodka. The media were informed about the plans to drop the flag before they had successfully reached the ocean bed, and Lavrov was informed about the expedition afterward. The Russians realized that they could use the record and all six were awarded medals by President Vladimir Putin. The three that got there first were given Russia's highest order: Hero of the Russian Federation.

But did *Mir-1* and *Mir-2* actually achieve what they claimed?

Five years later, on September 7, 2012, the German icebreaker RV *Polarstern* arrived at the North Pole. Its first task was a dive, straight down. When the submarine got to the seabed, the crew filmed an area of about three hundred feet. Looking out through the submarine windows, the Germans did not see a Russian flag, and no flag is visible in the film either.[31] However, the scientists on board *Polarstern* have done little to publicize this, unlike other polar explorers who have discovered something unexpected.

A recurring question in the history of the North Pole remains unanswered: What was the position of the Russian expedition when they left the flag on the seabed? The explorers were experienced navigators, with top equipment, and I believe they reached their goal. Paulsen said that the flag was made on board the icebreaker, after they had been drinking vodka and that it was no more sturdy than the flags that are put out on conference tables.[32] He laughed when he said this. In all likelihood, the flag could not withstand the pressure and ocean currents and had been swept away.

8

WHEN THE DREAM OF THE NORTH POLE WAS REALIZED

Erling between ice hummocks on his way to the North Pole.

The North Pole Is Not a Place

The North Pole is a fixed point on a globe. But on May 8, 1990, when both our GPS and dead reckoning showed that we were at the Pole, it felt as though it did not exist. The ice was the same as the day before, our surroundings were the same as the day before that: As far as the eye could see, white, gray, and gray-blue ice that was almost as flat as a sheet of paper in some places, and in others twisted up into hummocks several feet high—broken only by open leads. When I scrambled up onto the top of a hummock and looked out to the horizon, it was no different: ice suspended on the surface of the ocean, drifting on the currents, in every direction. We were at the North Pole, but at the same time, knew that we were already moving away from it. By the time we woke up the next morning, we would have drifted south.

The North Pole and the South Pole are the world's reductio ad absurdum. The two points have no longitude. The poles have no position. They are a nowhere. The North Pole is "just a thing you discover," as Christoper Robin says when Winnie the Pooh asks what the North Pole is. And when you get there you will find, as Winnie says, "Just something." A place that is in effect ice, that is constantly moving, breaking up, melting, and disappearing, and where the coordinates wipe out time and space, is not really *something*, but nor is it nothing. Which is what the Russians discovered in 1948. Their experience is a reminder that the North Pole is at the top of the world, but soon after you get there, you will have drifted away to somewhere else. Perhaps that is why it is a natural paradox that the explorers who were definitely there, in a way, both exist and do not exist—yet the twenty Russians are absent in most accounts of North Pole exploration.

It is often important to take something home from any journey of exploration over sea and land—a souvenir, something concrete, proof. But all there is at the North Pole is cold air, cold water, and ice. The South Pole explorer Robert Falcon Scott and the astronaut Neil Armstrong could at least take some stones. The North Pole is like an abstract painting, freed from all form. There is nothing of real substance there. And yet it is a destination, the start of many adventures, and a source of myths throughout recorded human time.

THE FIRST THING that Børge and I did when we got there was give each other a big, long hug. We had never embraced before. Then we attached the Norwegian flag to a ski and clambered to the top of a hummock and posed for a picture with a self-timer. We held the ski with both hands and smiled at the camera. A gentle gust of wind made the flag flutter ever so slightly.

We wondered if we were first. Could the British have gotten there before us, then drifted away again, so the polar point looked untouched by the time we got there?[1] It was impossible to know. Whatever the case, we had done our best.

Hunger gnawed at us all the way to the Pole—when we were skiing and when we were falling asleep. We had used up all the fat in our bodies and as we started to run out of things to talk about, all we spoke about was food. We had worn the same woolen underwear night and day, without taking it off, and when we got to the North Pole, Børge pulled his woolen vest up over his stomach and chest, almost all the way up to his neck. He looked like he was nothing more than loose, dirty skin and bones. His ribs were sticking out. I started to laugh. Not because it was funny, but because I had no other response left, only laughter. Børge asked me to pull up my vest. I saw my torso for the first time in fifty-eight days. I was just as emaciated. I ran my right hand down over my ribs, from my armpit to my buttocks. The muscles had vanished, I felt only bone and sinew.

When we were forced to shoot the polar bear at 88° 19′ N, Børge cut off some pieces of meat. He had learned to butcher a sheep when doing his military service, and was fairly confident he'd cut a tenderloin. We had dreamed of eating the meat every day since. As the guidelines were rather unclear about what exactly was meant by getting to the North Pole without dogs, depots, and motorized aids, we did not dare eat the bear meat until we got there. The explorer Ran Fiennes had mentioned to us that eating what you had hunted along the way could be seen as a form of support.[2] This seemed a little odd to us, but we did not want to chance it, no matter how hungry we were.

To avoid the risk of trichinella, we boiled the frozen meat, then fried it. We sat in silence and watched the meat boil, and were equally speechless when it was being fried. I then divided it into two portions, and Børge chose his first. When we finally got the meat in our mouths, it tasted like cod liver oil and was tough and chewy, but we did not care. I joked that Børge's experience of cutting up a sheep was hardly a guarantee that he could find the tenderloin in a polar bear. To get as many calories out of it as we could, we drank the cooking water for dessert.

A few years after I was given my first globe, I wondered what it must feel like to time travel. At the North Pole, it seemed quite natural. It only takes a few steps to walk around the world and cross twenty-four time zones. If you walk with the sun, in other words west, 360 degrees around the polar point, you walk into yesterday when you pass the dateline. If you then turn around and walk against the sun, or east, after circling once you step into the future.

Børge and I walked around the North Pole both ways. First west and then east. We thought it would be an exciting experience and had looked forward to it, and I would love to tell you something fascinating about our walk, but there is little to report. We were too tired to enjoy the fact that we were traveling through time, even though the idea was just as mind-blowing as it'd been when I was a boy.

We lay drifting by the North Pole and had no idea when we might be picked up. The midnight sun warmed the tent. It felt like we had come to paradise: due to the fact that it was not cold, and we could finally rest, sleep as long as we liked. A different paradise from the North Pole imagined by people throughout history. But everything that comes at a cost also gives the greatest pleasure, as Fridtjof Nansen wrote. To rest in a tent warmed by the midnight sun was the greatest luxury I could imagine, and then life got even better when the American spy plane dropped the food and questionable reading material.

"Even on the pack ice, everyone can find their slice of paradise," I wrote. After a few days, I started to take our paradise for granted, and the delight in being able to lie quietly in my sleeping bag slowly diminished. The tent was almost buried in snow on the second night as a strong wind swept through, and we had to go out and clear the snow. It was almost as though the storm brought us back to our senses.

The Twin Otter that was going to pick us up needed an airstrip that was flat enough for it to land safely, and long enough for it to take off. We used the axe and spade to prepare the airstrip and then compressed the wet ice with our skis. Whenever we took a break, we talked about Roald Amundsen and his five teammates in 1925. They had done exactly the same thing, only their situation was more precarious. The airplanes were not suited to land and take off on ice, and the likelihood of being rescued was close to zero for Amundsen's expedition.

Five days later, weather conditions between Canada and the North Pole were good enough for the Twin Otter to come and collect us. As soon as the plane landed, we rushed over to the pilot and he told us that the expeditions from the UK, Russia, South Korea, and Canada had all capitulated.

It had taken more than two years to prepare for the expedition. I had focused on learning about ice, ocean currents, weather conditions, and previous expeditions to help me understand what was needed to get to the top of the world. There was no time for anything else in my life. For the fifty-eight days it took to complete the expedition, I did not have the headspace to think about anything other than what was required to keep heading north. But in the course of those few days at the North Pole, I started to reflect on our achievement.

Directly above where I stood on May 8, my body in line with the imagined axis that runs through the earth from the South Pole to the North Pole, around which the world turns, was the celestial North Pole. The center of all the stars in the northern hemisphere felt like the mythological ground zero in polar history. The imaginary North Pole; the celestial North Pole; the geographic North Pole—the three poles would not exist had Ursa Major not moved eternally in orbit high up above the point where we now stood. And without the geographic North Pole, there would be no pole in the south.

Throughout human time, our ancestors in the northern hemisphere have wondered about the celestial North Pole, and tried to understand what was hidden in the stars of Ursa Major, and what was on the ground below the constellation. My feet on the ice, planted at ground zero, felt very real, unbelievably solid. And yet the North Pole was still a strangely abstract notion, a perfect source for mysteries and myths.

The sun blazed when we were at the North Pole. We did not see Ursa Major and the other stars that, in their own quiet way, since I was a child, always made me feel how vast and unexplored the world is, and how minute I am in all that vastness. I lay there in my sleeping bag and reflected on the unanswerable questions asked by our ancestors, and marveled at our ability to question, dream, and seek out adventure.

And from that moment, half a lifetime ago, I began the book you hold in your hands.

AFTERWORD: ICE, FATHERS, AND THE FUTURE

KVITØYA, SEPTEMBER 2023

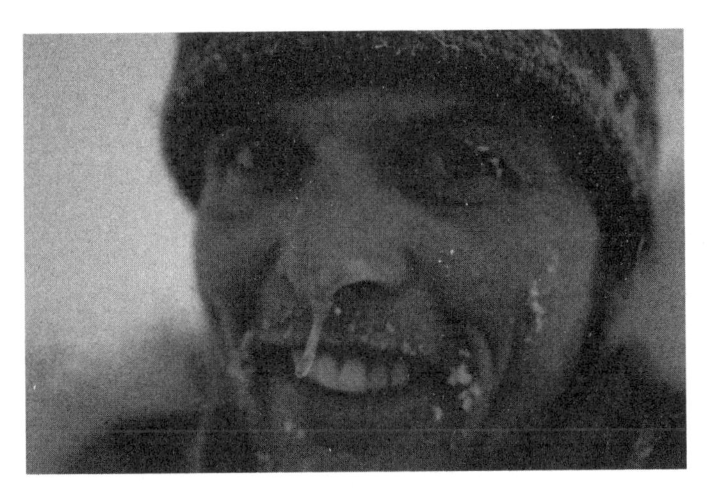

Erling, cold but smiling.

We stayed on Kvitøya for three days. The day after the polar bear had prevented us from looking around Andrée's base, Børge, Håvard Tjora, and I put on our skis. Håvard had lived on Svalbard for five years. And during that time he had tried to sail to Kvitøya twice, and both times had been stopped by the pack ice.

Our aim was to honor Andrée, Frænkel, and Strindberg with a ski trip—from Andréeneset in the southwest to the opposite coastal strip, Kræmerpynten, in the northeast. Until then, no one had traversed the island on skis.

As on previous expeditions, we took no more risks than were necessary. There is always a chance that everything will be fine and a chance that everything will go wrong—very wrong. Nothing is absolute. The degree of likelihood lies somewhere between 0.1 and 99.9 percent.[1] It would be natural to be paralyzed by the thought of all that could go wrong, but it is better to think that, no matter how difficult a situation is, there is always reason for hope.

We each pulled our own pulk, on which we had oats mixed with baby formula and some extra fat, dried meat, and fuel. We took enough for five days, even though the trip should only take two. We also had tents, sleeping bags and sleeping mats, rope in case there were crevasses, and two flare guns to frighten any polar bears that might get too close. And just in case, Håvard had a rifle and Børge and I had the two Magnum revolvers we had taken with us to the North Pole all those years ago.

Just before we set off into the mist that lay low over the island, we spotted a polar bear about three hundred feet away. The bear had already seen us and was looking in our direction. A bear can attack and kill even when it's not hungry, but this one stayed where it was and watched us for a few tense minutes. When we started to ski into the mist, it followed us with its eyes, but did not move. We never saw it again.

An Ice-Free Arctic Ocean

August Petermann spent his life trying to prove the theory of an ice-free Arctic Ocean, and countless polar explorers risked their lives believing it to be true. Eventually, however, it became clear how ridiculous this theory was. After the thirteen survivors of the *Jeannette* expedition returned in 1881 and described the pack ice that had eventually crushed their ship, the theory lost all traction. This was then followed by accounts from George Nares, Robert Peary, and Roald Amundsen, who all talked about an ocean covered in ice. The idea of an ice-free ocean was then seen to be an impossibility.

Geologists today believe that the temperature at the North Pole was 57–72°F warmer between 3.9 million and 5.8 million years ago. This would mean that summer temperatures could rise to more than 68°F at the Pole, and the Arctic Ocean was in fact ice-free for a few months every year. Farther south, along the coasts of what are today Canada, Russia, Alaska, and Greenland, the temperatures were even warmer.

The remains of camel bones have been found on Ellesmere Island—where it was so cold when we set off toward the North Pole that any particles containing smell froze to ice. The animals lived there more than three million years ago. Badger and horse bones have also been found in the same area. Paleontologists believe that camels, with their two humps, were well suited to survive in such an environment. The humps are basically fatty reserves that the animals could deplete through the winter until temperatures rose and things started to grow again in the Arctic.[2]

It now seems that the historians from Herodotus through to William F. Warren were right when they wrote about a warm climate around the North Pole. As was the freedom fighter Bal

Gangadhar Tilak, who believed he had found evidence of the same in the Hindu Rigveda—but the timing was incorrect. The Polar Ocean has been covered by ice for the last 2.7 million years. At that time the world was already cooling and a number of ice ages followed. We are currently between two ice ages. No one knows for certain why the cooling happened, but movements in the surface of the earth may have played a role. The ocean floor beneath the Arctic Ocean rose, so that it became shallower, and in the south Alaska and Russia were still joined. The mainland then sank into the ocean, creating the passage that was later named the Bering Strait. In the area between North and South America, close to what is now the Panama Canal, there was a natural passage from the Atlantic Ocean to the Pacific Ocean. When the ocean bed rose, the strait disappeared and the flow of water between the two oceans was restricted. The flow of fresh water into the Arctic Ocean from rivers in Siberia, combined with less movement and water circulation, resulted in favorable conditions for ice formation. There was more fresh water, which freezes at 32°F, and less salt water, which freezes at a little below 28.4°F.

Homo habilis, one of our earliest ancestors, appeared around 2.3 million years ago.[3] The cooling climate required new tools and they were one of the first species to use stone. It has taken around 75,000 generations for the species to evolve from *Homo habilis* to present-day *Homo sapiens*. And for each generation, the North Pole and Arctic Ocean have been covered with ice.[4]

Kvitøya, which is part of Svalbard, can today be used as a barometer for how the world is changing. In the past fifty years, the temperature on Svalbard has increased by as much as 7.2°F. One reason why the temperature is rising faster here than in other places is that the Barents Sea is now ice-free, so the sunlight is no longer reflected off the ice, and so is heating the dark ocean, causing a rise in the temperature on land and sea. The changes that are currently happening in the north may be a warning of how changing temperature will affect our environment and, thus, how we live. As Margaret Atwood warned, the earth, like trees, "dies from the top down. The things that are killing the North will kill, if left unchecked, everything else."[5]

THE MIST HUNG over Kvitøya through the first day and night of our expedition. The only color on the interior ice was white, on white. The ice was white, the air was white—and the polar bears were white. We kept looking over our shoulder for bears, and took turns in keeping watch while the others slept.

The mist lifted early in the afternoon of the second day. We had crossed the top of the island and could see the Arctic Ocean. To the far north of the horizon lay the North Pole. I got the feeling I was looking into a possible future Kvitøya. With the exception of a few icebergs, the ocean was ice-free. Somewhere out there to the north, Andrée, Frænkel, and Strindberg had struggled across the ice; to the east, the pack ice and cold had stopped Julius Payer and many other expeditions; and to the west, multiple American, British, and German expeditions had given up because the men did not stand a chance in the battle against the elements. Now it was open water.

After the Mechanical Age, a number of expeditions have succeeded in crossing the ice from Canada, Alaska, and Russia to the North Pole. The American Ralph Plaisted led a snowmobile expedition in 1968, which was the first to definitely reach the North Pole without flying.[6] Plaisted gave a good summary of what it takes to reach the North Pole: "Think ahead, travel light, and leave your fears behind you."

Wally Herbert's expedition got as far as the coast of Svalbard the year after Plaisted, having started from Alaska the year before. At the end of the 1980s, I rang him in connection with our planned expedition, and the advice he gave stayed with me—among other things, never take anything for granted on the ice. The breadth of his experience was impressive and it struck me that he, like so many other polar explorers, had never been given the recognition he was due.

The Japanese explorer Naomi Uemura (1941–84) was the first to go solo to the North Pole in 1978. He had supplies dropped to him from an airplane along the way. He did not give up, not even when he was woken one night by a polar bear sticking its head in through his tent flap and sniffing his sleeping bag. He thought about all the people who had helped him and that he "could never face them if I gave up." I know that thought only too well. The people I least want to disappoint are those who have made valuable contributions. In 1979, the Russian explorer Dmitry Shparo led the first expedition to ski to the North Pole. When Børge and I were finally at the North Pole ourselves, we wondered if, at some point in the future, someone might ski to and from the North Pole unaided. I wrote: "They have to hurry. If climate change continues, there will be less and less ice up here." Five years later, the Canadian Richard Weber and Russian Misha Malakhov succeeded in doing just that.

In 2019, Børge and Mike Horn crossed the Arctic Ocean in winter. They set off on their skis on September 11. The annual sunset over the Arctic had already started and on September 23 the sun sank below the horizon for the winter. Even though they could no longer see the sun, it continued to light up the days and nights for another three weeks. In the middle of October, there was darkness. Those polar explorers who have seen the ice shrinking with their own eyes now generally believe that in the future there may be so little ice that expeditions can no longer set off north from the coast.[7]

Perhaps August Petermann will be finally vindicated after 200 years. The temperature is rising in the Arctic at twice the speed of the global average. After 75,000 generations with an ice-covered Arctic Ocean, we may be the first generation to experience an ice-free ocean in summer.

IF A LARGE part of the Arctic Ocean becomes ice-free in summer, like the area we saw from the top of Kvitøya, there will be great changes. The warmth from the sun will be absorbed by the black water. The temperature will rise. Over time, polar bears might even change color, if they are able to survive at all close to the North Pole.

If I ever become a grandfather and want to give my grandchild a globe, it will no longer be like the one I grew up with. The pack ice around the North Pole will be considerably smaller and may even disappear for large parts of the year within the next generation. Then it will no longer be possible to hear the song sung by the ice for the past 2.7 million years.

A new imaginary North Pole is in the making. Europe, Asia, and North America are already closing in on the North Pole. New shipping routes will be opening up over the Arctic Ocean and the dream of the British and the Dutch from the 1500s to cross the Polar Ocean to the Pacific and Indian Oceans may become a reality in the summer season. Peter the Great's dream of an ice-free coastal route through the Northeast Passage, controlled by Russia, can also be realized. Territorial rights may also be about fishing and valuable minerals under the seabed and rights to airspace, and beyond that airspace, the space toward the celestial North Pole.

Today it is clear that the politicians and top-ranking military officers in Russia, the US, and Britain who believed the Arctic Ocean would be of crucial importance in the event of a Third World War were right to worry.

There is an arms race going on in the north, in the shadow of all the other conflicts that exist in the world today. Russia is reopening old bases and building thirty new ones all the way along the Northeast Passage, and the US, Canada, and NATO are reinforcing the Northwest Passage—for surveillance purposes, in a bid to stop an attack by the other and secure the shortest distance for an attack on the other. Two-thirds of the nuclear strike capabilities of the Russian Navy are stationed in the Arctic, on the Kola Peninsula, next to the Atlantic Ocean and forty miles from Norway. Missiles on board twenty-six submarines can reach most major cities in Europe and the US. Military experts in NATO agree that a possible Third World War would not start in the north but would quickly spread to the Arctic region. This is because Russia fully depends on the weapons in the Arctic to ensure second-strike capabilities in a nuclear war.

WHEN WE HAD crossed Kvitøya and were approaching the coast again, I stopped and looked out over the Arctic Ocean, which stretched to the North Pole and on, all the way to the Bering Strait. I knew that pack ice started just beyond the horizon. And in that moment, I remembered that the ice has never needed us—it is we who need the ice, now more than ever.

Fathers and Sons

We choose to do what we do for more than one reason, and while I am not suggesting that a polar explorer's motivation can be explained simply in terms of his relationship to his father, I do think it is one of the driving forces for most.

The British explorer Ben Saunders—who went solo to the North Pole in 2004—alerted me to an experience shared by a number of polar explorers, from Frobisher to Cook, Henson and Peary, Andrée, Amundsen, and himself. Their fathers died when they were children, or left the family and had no contact with their sons.

Ben's father left when Ben was ten years old. He left without saying goodbye, and never came back. Our competitor when we were skiing to the North Pole, Ran Fiennes, lost his father four months before he was born. The father of one of my role models, the British mountaineer Chris Bonington, left Chris when he was nine months old. William Ziegler lost his father when he was three, and Henry Stanley when he was only a few weeks old. In general, a difficult childhood may be tough, but it is a good start for an adventurer. I asked Ran if he thought he would have tried to reach the North Pole if his upbringing had been easy. He shook his head.[8]

In *The Odyssey*, which my mother read to me as a boy, Odysseus's son Telemachus asks himself: "How can I get to know my father better?" Or, to put it another way, who gives Telemachus the experience, knowledge, and confidence he needs when growing up, when his father is never home? What is it like to grow up with a father who goes to battle, fights monsters, outwits gods, and shares his life with goddesses, rather than with his son?

The lives of parents before having a family are a mystery to the child. For Telemachus, the next twenty years are also a mystery. *A son has always trouble at home when his father has gone away*. When staying at home is no longer an alternative, Telemachus goes out into the world. He has to find Odysseus. Find out if he is alive. Get to know him. "It is a wise child that knows his own father," he says to his mother, Penelope, before he sets off.[9] Telemachus sails over the ocean, as his father did, in search of an answer. I see the journey as more than just the search for his father; it is also a kind of substitute for the time and experiences they have not shared. Sometimes, when I meet a new polar explorer, I find myself thinking that this person is doing exactly what Telemachus did.

When I returned from the North Pole in 1990, my father said: "I don't know why you ever went to the North Pole. I always thought it was a ridiculous thing to do." And he added that if anyone should have gone on the expedition, it should have been my eldest brother, not me. In a way he was right: My brother was better at skiing, but he was not obsessed with the North Pole. Our father was, in his own way, a version of the age-old archetype—a father who found it difficult to praise his son.

It is easy to think your problems are unique. But I have come to understand that most boys have a difficult relationship with their father, and feel unsure of who he is. And most fathers struggle with their sons. Fortunately, since 1990 my father and I have grown closer and have a strong relationship. Today we are good friends. There are many reasons why people choose to go to the North Pole, and one of the reasons I became a polar explorer was to win my father's respect. Now that I am an adult, I have gradually come to understand that I share many of his traits—as Telemachus eventually discovers with his father—and I forgave him a long time ago, as I hope my children will forgive me for my *eventyrlyst*.

I WILL NEVER understand all the reasons why people want to go to the North Pole. I am not even sure of my own motives. A polar explorer has so many dreams, including, of course, doing something phenomenal, achieving something no one else has ever achieved. The certainty of frostbite, hunger, and exhaustion makes most people choose other more tolerable things when they decide to live out their dreams, or they give up almost as soon as they have started. But, as the explorer Ann Bancroft said, polar explorers also go to the North Pole precisely because of the uncertainty and the cold. When you are in the middle of nowhere, tapping on a phone screen does not give a sense of security or raise the temperature. For me, it is a relief not to live by screen alone. I can still remember what I thought when the Twin Otter took off on March 8, 1990, and left us standing on the ice in −62°F. The only thing that separated us from the pack ice was our .3-inch-thick skis. When it is that cold, you only have one thought in your head: It is cold. But when you are so close to nature, you experience something like gratitude. What gives rise to that feeling? Explorers find out the answer on the ice.

It is easy to overlook everyday feelings and experiences, and even harder to remember them, as the days and years pass. When we got back from Kvitøya, Børge was asked why he goes on polar expeditions, and he replied that it is not the days "lying on the sofa at home that you remember, it's the days when you had to fight, when it was cold and icy and you had profound experiences."

The German philosopher Arthur Schopenhauer once said: "Only pain and desire can truly be felt."[10] He was almost right. But he'd forgotten one important aspect: the pleasure you feel when the pain is still there, but you know it is about to loosen its grip. First there is the numbness, then you start to feel the warmth. In your chest first of all, and then it spreads all the way out to your toes, nose, ears, and fingertips. A slow warm wave of well-being. You feel your heart pumping the blood through the arteries and veins to your whole body. It is that experience, that one thought, that draws me back to nature.

MOONLIGHT VIEW OF THE "ROOSEVELT" IN HER WINTER QUARTERS
Exposed with full moon for three hours, December 12, 1905, by Dr. Wolf, Surgeon of the Expedition

The surgeon on Peary's North Pole expedition took this photo of the boat stuck in the ice in the light of a full moon, with a three-hour time elapse on December 12, 1905.

THANKS

With thanks to my editor in Oslo, Joakim Botten, and to Petter Skavlan, who has read the manuscript several times over the years and given me valuable advice, and to my English editor, Greg Clowes, and American editor, "Biz" Mitchell, for their inspiration and good help. Thanks also to my colleagues in Kagge Forlag: Cathrine Sandnes, Solveig Øye, Hans Petter Bakketeig, and Nora Dahle Borchgrevink, as well as Aslak Nore, Åsne Seierstad, Bjørn Gabrielsen, Ulrich Sonnenberg, Geir Kløver, Thorleif Thorleifsson, Gabi Gleichmann, Morten Magnus Faldaas, Edvard Hambro, and my mother, Aase Gjerdrum, who have all read the book and given me good and useful feedback.

My deepest thanks go to Børge Ousland and Geir Randby for their company on the journey north, and their crucial support for this book. And most importantly, without our shared dream, this book would never have been written. My thanks also to Børge for permission to use his photographs.

I deeply appreciated all the help and advice from Anders Bache, Victor Boyarsky, Noah Regenass, Christian de Marliave, Kristin Brandtsegg Johansen, Gary Fisketjon, Kari Dickson, Jonathan Landgrebe, Cecilie Skog, Nina Hovda Johannesen, Unn Falkeid, Lars Fr. H. Svendsen, Alexander Wisting, Anna Calame, Christian Katlein, Haraldur Örn Ólafsson, P. J. Capelotti, Charlotte Sabella, and Yonca Dervisoglu. And my thanks to my agents Zoë Pagnamenta at Calligraph, Annabel Merullo from PDF, and Hans Petter Bakketeig and Leyla Körner Øier from the Stilton Literary Agency.

Thanks also to my first publisher in the US, the late Sonny Mehta, who encouraged me to keep on writing. To Ranulph Fiennes for the competition in the 1990s and for our friendship since. And last, but not least, huge thanks to my English translator, Kari Dickson.

I am a deeply grateful author. Without their generous support, I could never have written this book.

NOTES

INTRODUCTION

1. A Nordic type of sled pulled by skiers.

CHAPTER 1: THE FOUR NORTH POLES

1. How many North Poles exist is up to each individual. Others have also written about the Arctic Pole of Inaccessibility, the point in the Arctic Ocean that is furthest from land. There is another North Pole inside the earth, directly below the geographic North Pole. I feel no particular relationship with these poles.
2. Research has shown that the axis that runs from the South Pole to the North Pole moved 1.8 inches south toward Canada annually between 1993 and the turn of the millennium. The axis has since moved eastward. Compare Ki-Weon Seo, Dongryeol Ryu, Jooyoung Eom et al., "Drift of Earth's Pole Confirms Groundwater Depletion as a Significant Contributor to Global Sea Level Rise 1993–2010," *Geophysical Research Letters* (June 2023), https://agupubs.online library.wiley.com/doi/10.1029/2023GL103509.
3. Job 9:9 and 38:32.
4. Wikipedia, "Ursa Major," https://en.wikipedia.org/w/index.php?title=Ursa _Major&oldid=1147279246.
5. Tariq Malik, "North Star Closer to Earth than Thought," Space.com, November 30, 2012, https://www.space.com/18717-north-star-distance-measurement .html.
6. Magnetic north lies close to the geographic South Pole, and vice-versa—magnetic south lies close to the geographic North Pole. The North Pole has that name because the compass points to the geographic north, in other words, magnetic south, as opposite poles attract. The compass magnet's south points to the geographic south and magnetic north.
7. Alan Buis, NASA's Jet Propulsion Laboratory, "Flip Flop: Why Variations in Earth's Magnetic Field Aren't Causing Today's Climate Change," Ask NASA Climate, August 3, 2021.
8. V. Hart, P. Nováková, E. P. Malkemper et al., "Dogs Are Sensitive to Small Variations of the Earth's Magnetic Field," *Frontiers in Zoology* 10, no. 80 (2013).
9. P. Hore and H. Mouritsen, "How Migrating Birds Use Quantum Effects to Navigate," *Scientific American*, April 1, 2022.

10. N. Davis, "Pole Position: Human Body Might Be Able to Pick Up on Earth's Magnetic Field," *Guardian*, March 18, 2019, https://www.theguardian.com/science/2019/mar/18/humans-earth-magnetic-field-magnetoreception.
11. Ibid.; Hart, Nováková, Malkemper et al., "Dogs Are Sensitive to Small Variations"; Abraham R. Liboff, "Why Are Living Things Sensitive to Weak Magnetic Fields?," *Electromagnetic Biology and Medicine* 33, no. 3 (2014): 241–45, DOI:10.3109/15368378.2013.809579; Chris Gayomali, "Dogs Might Poop in Line with the Earth's Magnetic Field," *The Week*, January 8, 2015, https://theweek.com/articles/453642/dogs-might-poop-line-earths-magnetic-field.
12. If you are currently in winter time, the time would be 11:00.
13. The story is in *Papillon* or the sequel, *Banco*. As an adult, I have learned that the author had mixed fact and fiction, but in this instance, he was on Devil's Island, as he describes; he did try to escape several times and eventually succeeded. A quarter of prisoners sent to the island either fell ill or died within the first year.
14. This type of fuel is called white gas in Canada, where we bought it.
15. I cannot remember the name, and I have not found the name of who it was.
16. Sigmund Freud wrote about the fact that everyone at some time or another dreams about being a hero, only I can't remember where.
17. Other sources claim that it was Vasco Núñez de Balboa who gave it the name, when he saw the ocean for the first time, having navigated through what is now Panama. What he saw was in fact Panama Bay, which according to Joseph Conrad has some of the calmest waters in the world.
18. Leif Amund Håland, "Rotter og vitaminer," *Stavanger Aftenblad*, August 25, 2001.
19. Hauk Wahl, Arne Saugstad, and Morten Stødle.
20. Paul Zweig, *The Adventurer: The Fate of Adventure in the Western World*, Akadine Press, New York, 1974.
21. Nowadays it is the Hanging Gardens of Babylon that are deemed to be a wonder of the world, not the wall. However, it is uncertain if the gardens ever existed, whereas the ruins of the wall are still there.
22. Carlo Rovelli, *Anaximander and the Birth of Science*, Riverhead Books, New York, 2023.
23. Bal Gangadhar Tilak, *The Arctic Home in the Vedas*, Arktos Media, Budapest, 2021 (first published 1903).
24. *Arbeiderbladet*, April 3, 1926.
25. I have garnered this information and that on Sarasvati below from my reading of the Rigveda. The reference to the seven stars, or "the Seven Sages," is on p. 214 of the *Rigveda: Selected Verses*, selected and translated from Sanskrit by Chayan Kumar Seal, first published in 2020, possibly self-published. Other information and quotes about the Rigveda are taken from Tilak's book, *The Arctic Home in the Vedas*.
26. Such as Purânas, Siddhântas, and various astronomical works.
27. According to Tilak, the idea that a year is equivalent to a long day and a long night for the gods is clear and consistent in later Indian works.
28. Tilak, *Arctic Home in the Vedas*.
29. The Truth and Reconciliation Commission of Canada, *Honouring the Truth,*

Reconciling for the Future: Summary of the Final Report of the Truth and Reconciliation Commission of Canada. Children were already being sent to boarding schools in the 1800s, but numbers increased dramatically in the years prior to 1930.

30. Tilak's ideas and theories were well known in India, but were of little importance elsewhere. He faced criticism in India and was accused of corroborating the colonial powers' idea that civilization came from the north. Any theories that the world was in part populated by people from the north fell out of favor later, with the growth of national socialism.
31. A number of other myths included in Herodotus and the Buddhist scripture Theravada are very similar to those in the Rigveda. One of my favorites is the mystic Uttarakuru Kingdom, which is central to both Hindu and Buddhist mythology. *Uttara* means "north"—North Kuru—and the kingdom lay to the north of the known country, Kurus, in North India. They were separated by a mountain range, which is usually referred to as Himalaya, but in Hindu texts Mount Kailash is often mentioned, and in Buddhist texts Mount Meru. The inhabitants of Kuru are at times described as being like people in the real world, and at other times as mystical.
32. Herodotus, *Histories*, Book 3, trans. Aubrey de Sélincourt, Penguin Classics, London, 2003.
33. Ibid., Book 4, pp. 13–36.
34. Tomislav Bilić, "Crates of Mallos and Pytheas of Massalia: Examples of Homeric Exegesis in Terms of Mathematical Geography," *Transactions of the American Philological Association* 142 (2012): 295–328.
35. Jerry Brotton, *A History of the World in Twelve Maps*, Penguin Books, London, 2014. See note 17.
36. Barry Cunliffe, *The Extraordinary Voyage of Pytheas the Greek*, Penguin Books, London, 2003.
37. He may well have been inspired by Homer and *The Odyssey.* Six days' sail from Telepylos there is a floating island where night and day are different from anywhere else.
38. Fridtjof Nansen, *In Northern Mists: Arctic Exploration in Early Times*, trans. Arthur G. Chater, vol. I, William Heinemann, London, 1911, ch. II.
39. "Eratosthenes and the Invention of Geography," *The History of Exploration* (podcast), episode 8, February 25, 2017.
40. Columbus wrote about this in a letter to Isabella and Ferdinand dated August 31, 1498. I read it in Carol Delaney, *Columbus and the Quest for Jerusalem: How Religion Drove the Voyages That Led to America*, Free Press, New York, 2011, but the letter is quoted in several other books.
41. Fridtjof Nansen, *Vitenskap og moral*, 1908; Erling Kagge, *Silence: In the Age of Noise*, Viking, New York, 2017.
42. Bettany Hughes, *The Seven Wonders of the Ancient World*, Weidenfeld & Nicolson, London, 2024.
43. Egyptologist Pål Steiner, University of Bergen. We corresponded in April 2023; Hazel Muir, "Pyramid precision," *New Scientist*, November 15, 2000, https://www.newscientist.com/article/dn174-pyramid-precision/; Tim Radford, "Pyramids Seen as Stairways to Heaven," *Guardian*, May 13, 2001, https://www

.theguardian.com/world/2001/may/14/humanities.highereducation; Anne Salleh, "Mystery of the Pyramids Solved with Stars," *ABC Science*, November 17, 2000, https://www.abc.net.au/science/articles/2000/11/17/213140.htm; EduBirdie, "The Astronomy of Ancient Egyptian," February 24, 2022, https://edubirdie.com/examples/the-astronomy-of-ancient-egyptian/.

44. I walked around Cairo for an entire night in 2007.
45. William F. Warren, *Paradise Found: The Cradle of the Human Race at the North Pole*, 1885.
46. The idea has become unpopular, as a few decades later the Nazis in Germany built part of their ideology on the belief that the Aryans came from the northern Arctic and, as a result of the harsh climate there and other reasons, were superior to all other races.
47. In the *Elder Eddas* and the *Younger Eddas*, which were both written in the 1200s, it says that the tree stood at the center of the world.
48. His full name was Yahya ibn Habash as-Suhrawardī.
49. Kari Vogt, "Sufisme," *Store norske leksikon*, updated December 14, 2021, https://snl.no/sufisme.
50. Henry Corbin was a professor at the Sorbonne and the University of Tehran.
51. Henry Corbin, *The Man of Light in Iranian Sufism*, trans. Nancy Pearson, Omega, Richmond, Virginia, 1994. The references to Corbin and Suhrawardi are taken from here. Corbin has been criticized for favoring Shia Islam over Sunni Islam, and being a Western man who tried to understand and share a culture to which he did not naturally belong.
52. Koran 24:35.
53. References to Kubra are taken from Corbin, *The Man of Light in Iranian Sufism*; details about Mount Qâf can be found at Wikipedia, "Mount Qaf," https://en.wikipedia.org/wiki/Mount_Qaf.
54. *The Kalevala*, trans. Keith Bosley, Oxford University Press, Oxford, 1989, p. 546.

CHAPTER 2: A REVOLUTION IN PERCEPTION

1. Petrarch, "The Ascent of Mont Ventoux," in *Epistolae familiares* (IV, 1), 1350.
2. Unn Falkeid, "Petrarch, Mont Ventoux and the Modern Self," *Forum Italicum* 43, no. 1 (2009): 5–28, DOI:10.1177/001458580904300101; dialogue with Falkeid. Translations into English by the author herself.
3. Roland Huntford, "Going to Extremes," *New York Times*, November 20, 1988.
4. Andreas Brekke, "Kartene som forsvant fra klasserommene," VG, February 12, 2019, https://www.vg.no/nyheter/meninger/i/yvAj4J/kartene-som-forsvant-fra-klasserommene.
5. I visited the University Library in Basel on June 15, 2022.
6. Andrew Taylor, *The World of Gerard Mercator: The Mapmaker Who Revolutionized Geography*, Walker Books, London, 2004.
7. Translations of the text on the map from Latin to English by the library.
8. Giraldus Cambrensis, *The Itinerary of Archbishop Baldwin through Wales*, J. M. Dent & Sons, London, 1912.
9. Noah Regenass explained the significance of the three words.

10. Joseph Conrad, "Geography and Some Explorers," *National Geographic*, 1924, XLV, 1–6.
11. Ibid.
12. Nicholas Crane, *Mercator: The Man Who Mapped the Planet*, Phoenix Paperbacks, 2003.
13. Ibid.
14. Clive Holland, *Arctic Exploration and Development c. 500 BC to 1915*, Fram Museum, Oslo, 2013 (first published 1994).
15. James Evans, *Merchant Adventurers: The Voyage of Discovery That Transformed Tudor England*, Weidendeld & Nicolson, London, 2013. Evans refers to Richard Haklyut's work several times, but does not give a specific reference here. It also says on the British Library's websites that the expedition never went ahead, https://blogs.bl.uk/magnificentmaps/2014/12/england-and-the-north-east-passage.html. The websites' source is Evans's book.
16. Richard Hakluyt, *The Original Writings and Correspondence of the Two Richard Hakluyts*, ed. E. G. R. Taylor, The Hakluyt Society, London, 2010 (1935).
17. Paul Strathern, *The Other Renaissance*, Atlantic Books, London, 2023. There are several books and articles on this expedition. Unfortunately, there are no first-hand accounts from the Russian fishermen who found them and the only source seems to be the Venetian ambassador—or agent—to London, Giovanni Michiel, in 1555. His version is based on what the Russian fishermen told the British men who in 1555 came to bring the ships and dead men back to London, and later told the details to him. See, for instance, https://collections.dartmouth.edu/arctica-beta/html/EA15-79.html. It is even unclear whether the fishermen found the boats in 1554, or a later year.
18. Strathern, *The Other Renaissance*.
19. Frederick William Beechey, *A Voyage of Discovery Towards the North Pole: Performed in His Majesty's Ships* Dorothea *and* Trent, *under the Command of Captain David Buchan R.N.*, Richard Bentley, London, 1818.
20. Fred Roots, "Why the North Pole Matters: An Important History of Challenges and Global Fascination," *Canadian Geographic*, May 14, 2017.
21. Andrea Pitzer, *Icebound*, Simon & Schuster, London, 2021.
22. I first sailed in the polar regions in 1987, to Antarctica, and in 2023 around the north side of Svalbard to Kvitøya.

CHAPTER 3: THE POWER OF THE UNKNOWN

1. According to Edmund Halley, the only plausible mathematical explanation for the movement of magnetic fields in the north was that the earth was hollow and could be entered at the poles.
2. Margaret Cavendish, *The Description of a New World, Called the Blazing-World*, 1666. The book is available on Gutenberg.org.
3. Håvard Tjora.
4. I read this somewhere, but unfortunately can no longer remember where. But based on a number of descriptions I have read about him, I think it is possibly true.

5. James Poskett, *Horizons*, Penguin, London, 2023. Poskett's source is Valentin Boss, *Newton and Russia*, Harvard Univ. Press, Cambridge, MA, 1972.
6. Hampton Sides, *The Wide, Wide Sea*, Doubleday, New York, 2024.
7. Wikipedia, "Daines Barrington," https://en.wikipedia.org/wiki/Daines_Barrington.
8. Sides, *Wide, Wide Sea*.
9. Cook treated people with respect on his first expedition, but on his last expedition he was at times brutal and ruthless in his dealings with the indigenous people.
10. Sides, *Wide, Wide Sea*.
11. Poskett, *Horizons*.
12. Edmund Burke, *A Philosophical Inquiry into the Origin of Our Ideas of the Sublime and Beautiful*, R. and J. Dodsley, London, 1757.
13. Immanuel Kant, *Critique of the Power of Judgement*, trans. P. Guyer and E. Matthews, Cambridge Univ. Press, Cambridge, 2000, p. 139.
14. Friedrich never went to the Arctic, but as I stood and contemplated his painting, I was convinced that he must have had better sources than the two art historians claim. The first was the sketches he made of ice on the Elbe River, near where he lived in Dresden. This may have taught him a lot, but in my experience the ice on a river is very different from the ice in his painting. Another source could be accounts from British polar explorers who had returned from the Arctic Ocean. A less mentioned possible third source could be the group of Norwegian landscape painters that Friedrich was part of; J. C. Dahl and Friedrich were close friends. They discussed their work and exchanged ideas, and Dahl inspired Friedrich to study and sketch the clouds; Johannes Grave, *Caspar David Friedrich*, Prestel, Munich, 2017.
15. Iain Boyd Whyte, *Beyond the Finite: The Sublime in Art and Science*, Oxford University Press, Oxford, 2011; Melvyn Bragg, Janet Todd, Annie Janowitz et al., "The Sublime," *In Our Time* (podcast), BBC.
16. Conversation with Geir Kløver, director of the Fram Museum.
17. Pierre Berton, *The Arctic Grail: The Quest for the North West Passage and the North Pole, 1818–1909*, Viking Penguin, New York, 1988.
18. The most powerful eruption in the world since the Fimbul Winter of 536, when Norway and Sweden experienced three winters "without summers in between," as Snorri wrote 700 years later in *The Younger Eddas*.
19. "1816: The Year Without Summer," *In Our Time* (podcast), BBC; Peter Frankopan, *The Earth Transformed: An Untold History*, Bloomsbury, London, 2023.
20. William Scoresby, *An Account of the Arctic Regions*, vol. 1, Archibald Constable and Co., Edinburgh, 1820, p. 284.
21. Michael Bravo, *North Pole*, Reaktion Books, London, 2019; see also Scoresby, *An Account of the Arctic Regions*, 2 vols.
22. Fergus Fleming, *Barrow's Boys: A Stirring Story of Daring, Fortitude, and Outright Lunacy*, Grove Press, New York, 2001.
23. Also in indirect reference: "Who had darted that stone lance?" Scoresby, *An Account of the Arctic Regions*; Herman Melville, *Moby-Dick*, Richard Bentley, London, 1851.

24. Beau Riffenburgh, *The Myth of the Explorer*, Oxford University Press, Oxford, 1994.
25. Kerstin Knopf, "The Exquisite Horror of Their Reality: Native and 'White' Cannibals in American and Canadian Historiography and Literature," in Annekatrin Metz, Markus M. Müller, and Lutz Schowalter, eds., *Feasting Fitness? Cultural Images, Social Practices, and Histories of Food and Health*, WVT, Trier, 2013, pp. 19–46.
26. Fleming, *Barrow's Boys*, p. 37.
27. Berton, *Arctic Grail.*
28. Fleming, *Barrow's Boys*; Berton, *Arctic Grail.*
29. In the "Minik and the Meteorites," *The Quest for the North Pole* (podcast), episode 2, it was calculated that the calorie intake was only 1,774 per day.
30. Nutrition experts Halvor Holm and Per Kristian Opstad planned our food together with Børge, where fat was the primary source of energy.
31. Erling Kagge, *Walking: One Step at a Time*, Viking Penguin, New York, 2019.
32. Kenn Harper, *Give Me My Father's Body: The Life of Minik, the New York Eskimo*, Steerforth Press, South Royalton, VT, 2000. See also Fleming, *Barrow's Boys.*
33. Edward Parry, *Three Voyages for the North West Passage from the Atlantic to the Pacific, and a Narrative of an Attempt to Reach the North Pole*, Harper & Brothers, New York, 1844.
34. Berton, *Arctic Grail.*
35. Roald Amundsen, *My Life as an Explorer*, Cambridge University Press, Cambridge, 1927, pp. 2–3.
36. Robert Falcon Scott, *Scott's Last Expedition: The Journals of Captain R. F. Scott*, Pan Books, London, 2003, p. 464.
37. Fridtjof Nansen, *In Northern Mists: Arctic Exploration in Early Times*, trans. Arthur G. Chater, vol. 1, William Heinemann, London, 1911, p. 4.
38. My sources here are the polar explorers themselves, in conversation or by email correspondence.

CHAPTER 4: THE RACE TO REACH THE NORTH POLE

1. Charles Dickens (ed.), *Household Words*, December 1854; David Roberts, "Last Words Missing: The Mystery of Sir John Franklin and Polar History's Greatest Catastrophe," *National Geographic*, March 30, 2021.
2. Wikipedia, "Transatlantic Telegraph Cable," https://en.wikipedia.org/wiki/Transatlantic_telegraph_cable.
3. Beau Riffenburgh, *The Myth of the Explorer*, Oxford University Press, Oxford, 1994.
4. William Morton (1819–68), dentist and physicist.
5. Berton, *Arctic Grail.*
6. Pierre Berton, *The Arctic Grail*, Doubleday Canada, Toronto, 2001, p. 258.
7. Nansen, *Farthest North.*
8. Berton, *Arctic Grail.*
9. Ibid.
10. Bruce Henderson, *Fatal North: Murder and Survival on the First North Pole Expedition*, BruceHendersonBooks, Menlo Park, California, 2014.

11. Ibid.
12. Emil Bessels, *Polaris: The Chief Scientist's Recollections of the American North Pole Expedition, 1871–73*, trans. Will Barr, University of Calgary Press, Calgary, 2016.
13. Ibid.
14. Ibid.
15. Henderson, *Fatal North*.
16. Edward S. Cooper, *Vinnie Ream: An American Sculptor*, Academy Chicago Publishers, Chicago, 2009, and "A Motive for the Murder of Charles Francis Hall," an article by Russel A. Potter on the website Visions of the North, 2015, https://visionsnorth.blogspot.com/2015/07/a-motive-for-murder-of-charles-francis.html. Potter came across an envelope at an online auction that was addressed to: Miss Vinnie Ream, 726 Broadway, New York. The letter was stamped *U.S. Steamship «Polaris»*, and Hall had written his name on the envelope as sender. The letter was sent from Greenland, just before their departure.
17. William Barr, "Epilogue: Motive for Murder," in Bessels, *Polaris*.
18. Hampton Sides, *In the Kingdom of Ice*, Doubleday, New York, 2014. I wrote about this in *Philosophy for Polar Explorers: What They Don't Teach You in School*, Pushkin Press, London, 2006; reissued in a revised edition by Penguin Books, 2019.
19. Julius Payer, *New Lands within the Arctic Circle*, De Appleton & Company, New York, 1877.
20. Christoph Ransmayr, *The Terrors of Ice and Darkness*, trans. John E. Woods, Grove Press, New York, 1991.
21. Payer, *New Lands*.
22. Fergus Fleming, *Ninety Degrees North*, Grove Press, New York, 2001.
23. Ransmayr, *Terrors of Ice and Darkness*.
24. Ibid.
25. Ellen Gutoskey, "The Turning Point," *The Quest for the North Pole* (podcast), episode 3, July 27, 2022, https://www.mentalfloss.com/article/640878/quest-for-north-pole-episode-3-podcast-transcript. The quotation in the podcast is taken from the *Pall Mall Gazette*.
26. Berton, *Arctic Grail*.

CHAPTER 5: THEORY AND REALITY

1. Vilhjalmur Stefansson, in the introduction to Alden Todd, *Abandoned: The Story of the Greely Arctic Expedition 1881–1884*, University of Alaska Press, Fairbanks, 2001 (first published 1961).
2. Glenn M. Stein, "An Arctic Execution: Private Charles B. Henry of the United States Lady Franklin Bay Expedition 1881–84," *Arctic* 64, no. 4 (2011), DOI:10.14430/arctic4139.
3. Javier Cacho, "Arctic Obsession Drove Explorers to Seek the North Pole," *National Geographic*, January 24, 2020. The source of the quotation is not given in the article; it is something that Greely "reportedly said."
4. Todd, *Abandoned*, p. 285.

5. Sides, *In the Kingdom of Ice.*
6. Martin Dugard, "Stanley Meets Livingstone," *Smithsonian Magazine*, October 2003, https://www.smithsonianmag.com/history/stanley-meets-livingstone-91118102/.
7. Sides, *Kingdom of Ice*, p. 5.
8. L. F. Guttridge, *Icebound: The Jeannette Expedition's Quest for the North Pole*, Naval Institute Press, Annapolis, Maryland, 1986. The quote is cited slightly differently in Sides, *Kingdom of Ice*: "The polar virus was in George's blood to stay."
9. Sides, *Kingdom of Ice*, pp. 174–75.
10. Others have claimed that De Long did not prioritize the officers.
11. Sides, *Kingdom of Ice*, p. 402.
12. Riffenburgh, *Myth of the Explorer*, p. 76. Sides, *Kingdom of Ice*, has a slightly different version.
13. Sides, *Kingdom of Ice.*
14. Ibid.
15. The first time I read a similar expression was on a drawing by Fredrik Stabell.
16. P. J. Capelotti, *The Greatest Show in the Arctic*, University of Oklahoma Press, Norman, 2016, p. 514; "Paul Bjørvig," *Norsk polarhistorie*, Frammuseet, Oslo, 2004, https://polarhistorie.no/personer/Bjorvig,%20Paul.html.
17. Riffenburgh, *Myth of the Explorer.*
18. Roland Huntford, *Nansen: The Explorer as Hero*, Barnes Noble Books, New York, 1998, p. 375.
19. Capelotti, *Greatest Show in the Arctic*, p. 139.
20. My source for what happened with Paul Bjørvig and Bernt Bentsen, and the quotes, is the handwritten diary of Bjørvig. On a few occasions the handwriting was difficult to read and I used two transcribed versions: "Paul Bjørvig," *Norsk polarhistorie*, Frammuseet, Oslo, 2004, https://polarhistorie.no/personer/Bjorvig,%20Paul.html and a version transcribed by Hans Aas of Vågemot forlag.

    Ten years after the expedition, Bjørvig returned to Norway after another winter in the Arctic. He was philosophizing about the men he knew who had died on expeditions, and wrote: "But if you have no sorrows, you also have no joys." Around the same time, Walter Wellman invited him on one more attempt to reach the North Pole, but Bjørvig said no. He had seen enough suffering.
21. This part of Pierre's life is told in *War and Peace*, Book Four, Part Three, chapters 12, 13, 14, and 15.

CHAPTER 6: THE HEROIC ERA

1. Eric Utne (ed.), *Brenda, My Darling: The Love Letters of Fridtjof Nansen to Brenda Ueland*, UTNE Institute Inc., 2011. The book is largely comprised of love letters from Fridtjof Nansen to his last great love, with photographs. Nansen's appearance would indicate that the photographs were taken fifteen to twenty-five years earlier.
2. Erling Kagge (ed.), *Eventyrlyst*, Kagge Forlag, Oslo, 2011. *Eventyrlyst* is a

collection of Nansen's speeches and articles. Originally published in Nordahl Rolfsen, *Illustreret Tidende for Børn*, Cammermeyers Forlag, Alb, 1891.

3. Fridtjof Nansen mentioned the quote several times. Carlyle wrote it in the essay "Goethe's Helena," published in *Critical and Miscellaneous Essays* (1838).
4. Fridtjof Nansen, "Mod Nordpolen." in Kagge (ed.), *Eventyrlyst*; originally published in Rolfsen, *Illustreret Tidende for Børn*.
5. Huntford, *Fridtjof Nansen*, p. 237.
6. Ibid., p. 187.
7. Nansen, *In Northern Mists*, p. 3.
8. Huntford, *Fridtjof Hansen*, p. 2.
9. Carl Emil Vogt, "'Et ikke ubetydelig bidrag': Fridtjof Nansens hjelpearbeid i Russland og Ukraina, 1921–1923," *Forsvarsstudier* 2 (2002).
10. The only exception I know of could be the *Jeannette* expedition, as it is unclear to me how much was actually planned.
11. Jon Sørensen, *Fridtjof Nansens saga*, Jacob Dybwads forlag, Oslo, 1931.
12. Nansen, *Eventyrlyst*.
13. Ibid.
14. This was the standard condition everyone on the *Fram* had to accept in their contracts. In his employment contract dated April 14, 1893, it first says "sailor." This has then been deleted and replaced by "stoker," which in turn was deleted and replaced by "participant."
15. Nansen, *Farthest North*, p. 75.
16. Several expeditions attached a small wheel to the back of the sledge that counted the number of feet covered.
17. Hjalmar Johansen, *With Nansen in the North: A Record of the* Fram *Expedition in 1893–96*, trans. H. L. Brækstad, Ward, Lock & Co., London, 1899, ch. XX.
18. Nansen, *Farthest North*, p. 449.
19. Erling Kagge, *Silence: In the Age of Noise*, Penguin Books, London, 2018.
20. Johansen, *With Nansen in the North*, p. 286.
21. The idea for the stove was Børge's, inspired by the stove that Nansen had made about a hundred years earlier.
22. They both wrote entries in their diaries on January 8.
23. Fridtjof Nansen, "Hjalmar Johansen: Some Memories," in Kagge (ed.), *Eventyrlyst*.
24. Huntford, *Nansen*, p. 340.
25. Ibid., p. 341. Jackson noted down their conversation shortly after, so the wording is likely to be very close to what was said.
26. Johansen's diary.
27. Fridtjof Nansen, "Speech for the Students of St Andrews University," November 3, 1926.
28. The quotes are from Nansen's article "På ski over fjellet. Fra Bergen til Kristiania," an account of skiing over the mountains from Bergen to Kristiania (now Oslo). The article was first published in *Aftenposten* in 1884, in Kagge (ed.), *Eventyrlyst*.
29. Vogt, "'Et ikke ubetydelig bidrag.'"
30. Walter Wellman also planned to fly in a balloon from Virgohamna to the North Pole, but did not manage to take off.

31. Alec Wilkinson, "The Ice Balloon," *New Yorker*, April 12, 2010.
32. Håkan Jorikson, *S. A. Andrée Ingenjören och polarfararen*, Carlsson, Stockholm, 2022. I have also had several email conversations with Jorikson about Andrée.
33. Nansen, "Speech for the Students of St Andrews University."
34. Kagge, *Philosophy for Polar Explorers*.
35. Günther Sollinger, *S. A. Andrée: The Beginning of Polar Aviation 1895–1897*, Russian Academy of Sciences, Moscow, 2005, p. 525.
36. Diaries on polar expeditions were often written for posterity, and Scott's diaries were edited with respect for the deceased.
37. Bea Uusma, *The Expedition: Solving the Mystery of a Polar Tragedy*, Head of Zeus, London, 2014. This book is also published with the title *The Expedition: A Love Story*. The following quotations are also from this book.
38. Jorikson, *S. A. Andrée*.
39. *Tidens Tegn*, November 2, 1935.
40. Nansen, "Speech for the Students of St Andrews University."
41. In addition to the references mentioned, the two following articles have been useful: Alec Wilkinson, "The Ice Balloon: A Doomed Journey in the Arctic," *New Yorker*, April 12, 2010, and Colin Dickey, "Above the Ice," *Paris Review*, October 23, 2014.
42. Robert Peary, *The North Pole*, Frederick A. Stokes, New York, 1910, p. 296.
43. Frederick A. Cook, *My Attainment of the Pole*, Mitchell Kennerley, New York & London, 1913, p. 23.
44. According to Cook, the name means "a windy place" (ibid., p. 69), and one of the advantages of a lot of wind is that it blows the snow away, allowing reindeer and musk ox to graze. According to others, *Annoatok* means "the wind-loved place."
45. I first read this in the *Los Angeles Times*, but have since read it in several other places.
46. Wally Herbert, *The Noose of Laurels: Robert E. Peary and the Race to the North Pole*, Atheneum, New York, 1989, p. 65.
47. Cook, *My Attainment of the Pole*, p. 27.
48. Ibid.
49. Jon Gertner, *The Ice at the End of the World: An Epic Journey into Greenland's Buried Past and Our Perilous Future*, Random House, New York, 2019.
50. Ibid.; Robert Peary, *Northward over the "Great Ice": A Narrative of Life and Work along the Shores and upon the Interior Ice-Cap of Northern Greenland in the Years 1886 and 1891–1897*, Methuen, London, 1898.
51. Berton, *Arctic Grail*, p. 525.
52. The three men said the same thing in slightly different ways.
53. As told by Siorapaluk in 1967, in Kenn Harper's *Give Me My Father's Body*.
54. Alexander Wisting, *Otto Sverdrup: Skyggelandet. En biografi*, Kagge Forlag, Oslo, 2017.
55. Otto Sverdrup, from Otto Sverdrup's diaries, cited in Wisting, *Otto Sverdrup*, p. 223. Sverdrup mapped 100,000 square miles on the expedition.
56. Robert M. Bryce, in the introduction to Josephine Peary, *My Arctic Journal: A Year among Ice-Fields and Eskimos*, new ed., Cooper Square Press, Lanham, MD, 2002.

57. Peary, *The North Pole*, p. 346.
58. David Welky, *A Wretched and Precarious Situation: In Search of the Last Arctic Frontier*, W. W. Norton & Company, New York, 2016.
59. Cook, *My Attainment of the Pole*, pp. 27, 43.
60. Bravo, *North Pole*.
61. In January 1990, *National Geographic* magazine published a new argument that Peary's expedition had indeed reached the North Pole, basing it on photographs and water depth measurements taken on the way. The photographs are of limited value as evidence, as much of the Arctic Ocean looks the same. The shadows on the photographs that Peary claims were taken at the North Pole are analyzed. The conclusion is that the shadows indicate that they may have been at the North Pole on April 6, 1909, as do the water depth measurements, according to the magazine. Two science periodicals, *Scientific American* and *Nature*, contested the results.
62. Peary, *The North Pole*, p. viii.
63. Cook, *My Attainment of the Pole*, pp. 6, 15, 517.
64. Michael Robinson, "Manliness and Exploration: The Discovery of the North Pole," *Osiris* 30, no. 1 (2015).
65. Herbert had himself crossed the Arctic Ocean—together with Ken Hedges, Roy "Fritz" Koerner, and Allan Gill—via the longest possible route, from Alaska to Svalbard, over the North Pole. The expedition was completed on May 29, 1969, having covered 3,720 miles over 476 days on the ice. When Herbert finally returned home to England, his neighbor asked where he had been. Apollo 11 had landed on the moon three weeks before, and the world's eyes were looking up, rather than north. If it could be proved that Cook and Peary never did get to the North Pole, then Herbert's expedition would be the first to reach the Pole on foot. It is therefore only natural to question how objective his conclusions might be.
66. Kenn Harper, "Liars and Gentlemen," The Frederick A. Cook Society, International Symposium on the Centennial of the Discovery of the North Pole, April 2008. One of the sources that Harper uses on several occasions is Andrew Freeman's book *The Case for Doctor Cook*, Coward-McCann, New York, 1961.
67. This story has previously been included in two of my books: Erling Kagge, *Nordpolen: det siste kappløpet*, Cappelen forlag, Oslo, 1990, and Kagge, *Silence: In the Age of Noise*.
68. Berton, *Arctic Grail*, p. 554.
69. Ibid., p. 578, where he quotes from an interview with Bartlett in the *New York Herald*, September 1909.
70. US Customs and Border Protection. "Did You Know a Customs Employee Was the 'First Man to Sit on Top of the World'?," last modified December 20, 2019, https://www.cbp.gov/about/history/did-you-know/first-man; Matthew Henson, *A Journey for the Ages: Matthew Henson and Robert Peary's Historic North Pole Expedition*, Skyhorse, New York, 2016. The first edition of the book was called *A Negro Explorer at the North Pole*, and in the foreword Booker T. Washington writes about Henson's claim that he was first to the North Pole.

71. I met Jamling Norgay in Geneva, June 18–20, 2023, in connection with the seventieth anniversary of Hillary and Norgay reaching the summit of Mount Everest.
72. Henson, *Journey for the Ages*, pp. 130–31.
73. Cook, *My Attainment of the Pole*, pp. 408, 415, 430.

CHAPTER 7: THE MECHANICAL AGE

1. In an article, "Utforskning av de ukjente nordpolare strøk," written in 1925, Fridtjof Nansen explains his idea to fly to the North Pole and explore unknown parts of the Arctic Ocean in an airship.
2. Beekman H. Pool, *Polar Extremes: The World of Lincoln Ellsworth*, University of Alaska Press, Fairbanks, 2002.
3. Ibid., and a conversation with Rita Ringnes, who knew the widow. She confirmed that Ellsworth's wife had told her the same.
4. Ibid.
5. Ibid.
6. Ibid.
7. Alexander Wisting, *Roald Amundsen: Det største eventyret*, Kagge Forlag, Oslo, 2011.
8. *Arbeiderbladet*, April 3, 1926.
9. Harald Dag Jølle, *Nansen: Oppdageren*, Gyldendal, Oslo, 2011.
10. These figures are given in several places, but without a source; and based on the accounts of Arctic expeditions that I have read, I would say the figures are believable.
11. Malcolm W. Browne, "Going to Extremes," *New York Times*, June 3, 1990.
12. According to Geir Kløver, director of the Fram Museum, Amundsen questioned Byrd's claim almost as soon as Byrd returned, and told an *Aftenposten* journalist: "He's not been there."
13. Wisting, *Roald Amundsen*.
14. *Ukens nytt*, May 20, 1926. I first read about this in Alexander Wisting's biography of Roald Amundsen.
15. *Stjørdalingen*, August 5, 1926. Likewise, I first read about this in Wisting's Amundsen biography.
16. Roald Amundsen, *My Life as an Explorer*, Cambridge University Press, Cambridge, 1927, p. 71.
17. Ibid., p. 72.
18. The flying boat had a crew of four men. The captain, René Guilbaud, was thirty-eight and one of the most experienced pilots in France. He had been preparing to fly a Latham over the Atlantic Ocean when he received the order to go north with Roald Amundsen. The Frenchman expressed his delight at the opportunity to fly with the famous polar explorer. The copilot was the thirty-year-old Albert de Cuverville. Like Guilbaud, he was a decorated war hero from the First World War, and had ample experience of flying a Latham. The twenty-six-year-old Gilbert Georges Brazy was known to be a skilled aviation mechanic who could fix anything. And finally, Emile Valette, twenty-eight

years old, had risen rapidly through naval ranks as a telegraphist. In the photographs taken in Bergen, they all look confident and happy.

19. Published in Italian in 1928 and in English in 1929.
20. "Lynsjestemning på Fagerneskaia–da Umberto kom til Narvik," *Wiker5's Blog*, June 15, 2015.
21. Peter W. Zapffe, "Roald Amundsens vei," *Aftenposten*, September 12, 1961, in connection with the fiftieth anniversary of his expedition to the South Pole.
22. Arne Næss, the Norwegian philosopher, told me this on one of our trips to his cabin. It must have been in the late 1990s, but I don't remember exactly when.
23. Sigmund Setreng, "Ambolten hvorpå selv gudene hamrer forgjeves," *Norsk filosofisk tidsskrift* 4 (1989), pp. 225–54.
24. I also learned a lot from Zapffe, "Roald Amundsens vei." See as well Peter W. Zapffe, *Kulterelt nødverge*, Pax, Oslo, 1997.
25. In the book *Unlikely Heroes: The Story of the First Men Who Stood at the North Pole* (Brussels, 1999), the author, Christopher Pala, writes about the Russians who reached the North Pole on this day. He does not mention all names, but writes that there were twenty-four men. In 1994, the last survivor from the expedition, Pavel Seniko, was interviewed about his days at the North Pole. Seniko does not mention the number of participants and he only uses surnames of several of the participants.

    In Russian sources, I have not found a comprehensive overview of everyone who was at the North Pole on April 23, 1948.

    The French polar explorer Christian de Marliave has worked to obtain a list with full names. He has managed to find all surnames, but not all given names. De Marliave's conclusion is that there were twenty Russians, and not twenty-four, at the North Pole on April 23, 1948:

    The leader of the expedition was Alexander Kuznetsov, and the captains on the three LI 2 planes were Ivan Cherevichny, Vitali Maslennikov, and Ilya Kotov. Vladimir Padalko, Morozov Chernusov, and Yuri Chernusov were the navigators. There were four scientists: Pavel Gordiyenko, Pavel Seniko, Mikhaïl Somov, and Mikhaïl Ostrekin. Two veterans from previous Arctic expeditions were also present: Ilya Masuruk and Mikhail Vodopianov. The radio operator was Cherpakov, and the mechanic Kekuchev. There were also three with unknown roles: Nikolaï Peskarev, Zaytsev, and Sokolov. In addition, the journalist Savva Morosov and cameraman Vladimir Frolenko were present.
26. Pala, *Unlikely Heroes*, p. 399.
27. Ibid., p. 340.
28. The polar explorers Victor Boyarsky and Christian de Marliave have helped me with information about the expedition.
29. Even in the detailed entry for Kuznetsov on the Russian website warheroes.ru, there is little to indicate that he possibly led the first expedition that, without any doubt, reached the North Pole.
30. We spoke on the telephone on April 26, 2024.
31. Conversation with Christian Katlein, from the Alfred Wegener Institute in Bremerhaven, at a private dinner on the deck of *Fram* in November 2023 and

subsequent email correspondence. The film is available on the Institute's website.

32. I did talk with Mike McDowell on the phone on September 9, 2024. Mike confirmed that he too was unaware of the geopolitical dimension of the expedition. As for the Russian flag, he believes it was made in Kaliningrad prior to their departure.

CHAPTER 8: WHEN THE DREAM OF THE NORTH POLE WAS REALIZED

1. The *Guinness Book of Records* and *National Geographic* magazine have both concluded, after having considered all other relevant expeditions, that our expedition was the first to reach the North Pole unsupported.

   Ran Fiennes claimed in 1990 that we had had support, because Geir was collected by an airplane, despite the fact that he had his gear, food, and fuel with him on the plane. When working on this book, I became aware that a Russian North Pole expedition on foot in 1989 has been written about as "unassisted." One man died and five were rescued by airplane. I have been informed that the men who remained kept the food and equipment of those who were flown out, and ate the food and used the fuel.
2. In a conversation Geir Randby and I had with Ran in his office at Occidental Petroleum in London. I can't remember the exact date, but it was probably in autumn 1988.

AFTERWORD: ICE, FATHERS, AND THE FUTURE

1. It was the philosopher Arne Næss who first spoke to me about the idea that nothing is absolute on an expedition, or in life—as long as one is alive.
2. Reidar Müller, *Ild og is*, H. Aschehoug, Oslo, 2022.
3. *Homo habilis* lived from 2.33 million years ago to 1.4 million years ago.
4. Eric Holthaus, "The Last Time the Arctic Was Ice-Free in the Summer, Modern Humans Didn't Exist," *Slate*, December 2, 2014. Holthaus writes that the top of the world has been covered by ice for the past 100,000 generations, and that, for the first time in three million years, the ice may disappear this century.
5. Margaret Atwood, *Strange Things: The Malevolent North in Canadian Literature*, Clarendon Press, 1995, Oxford, p. 140.
6. The three other members of the expedition were Walt Pederson, Gerry Pitzl, and Jean-Luc Bombardier.
7. In an article published by the Bjerknes Centre for Climate Research, "Den krympende sjøisen I Arktis," August 30, 2022, it is pointed out that while there are natural variations in the climate from year to year, century, and millennia, there are few observations that go so far back in time. "With regard to the shrinking sea ice, it is therefore not easy to separate the contribution of natural variations from the effects of climate change."

8. Ran and I had a few drinks together in Punta Arenas in autumn 1992. We were both about to go to the South Pole and were waiting for an airplane that could take us to Antarctica, so the mood was pleasant, albeit a little tense. I can no longer remember exactly what was said, but I remember the point.
9. Homer, *The Odyssey*, Book 1.
10. The Arthur Schopenhauer quotation on pain is taken from chapter 46 of *Die Welt als Wille und Vorstellung*, first published 1818. I read it first in Steven M. Cahn and Christine Vitrano (eds.), *Happiness: Classic and Contemporary Readings in Philosophy*, Oxford University Press, Oxford, 2008.

BIBLIOGRAPHY

A print showing the starting station in Danskøya for Swedish explorer Salomon August Andrée's tragic attempt to reach the North Pole by balloon.

In the time before and after our expedition to the North Pole, I have read books about expeditions north, by boat and on foot. My reading has also included books that might help to explain what these long and dangerous expeditions are about. I no longer remember all the books that I've read, but here are some that I have particularly enjoyed.

Aristotle, *Metaphysics*, trans. W. D. Ross, Clarendon Press, Oxford, 1924.

Astrup, Eivind, *Blant Nordpolens naboer*, H. Aschehoug, Kristiania, 1895; republished Kagge Forlag, Oslo, 2004.

Beechey, Frederick William, *A Voyage of Discovery toward the North Pole: Performed in His Majesty's Ships* Dorothea *and* Trent, *under the Command of Captain David Buchan, R.N., 1818*, Richard Bentley, London, 1843.

Berton, Pierre, *The Arctic Grail: The Quest for the North West Passage and the North Pole, 1818–1909*, Viking Penguin, New York, 1988. I have used this edition and the paperback edition published by Anchor Canada, Toronto, 2001.

Bomann-Larsen, Tor, *Roald Amundsen*, trans. Ingrid Christophersen, History Press, Cheltenham, 2006.

Bonington, Chris, *Quest for Adventure*, Hodder & Stoughton, London, 1981.

Bravo, Michael, *North Pole: Nature and Culture*, Reaktion Books, London, 2019.

Brotton, Jerry, *A History of the World in Twelve Maps*, Penguin Books, New York, 2014.

Capelotti, P. J., *The Greatest Show in the Arctic*, University of Oklahoma Press, Norman, 2016.

Cook, Frederick A., *My Attainment of the Pole*, Mitchell Kennerley, New York and London, 1911.

Cunliffe, Barry, *The Extraordinary Voyage of Pytheas the Greek*, Penguin Books, London, 2003.

Fernández-Armesto, Felipe, *Straits: Beyond the Myth of Magellan*, University of California Press, Oakland, 2022.

Figes, Orlando, *The Europeans: Three Lives and the Making of a Cosmopolitan Culture*, Penguin Books, London, 2020.

Gertner, Jon, *The Ice at the End of the World*, Random House, New York, 2019.

Harper, Kenn, *Give Me My Father's Body: The Life of Minik, the New York Eskimo*, Steerforth Press, South Royalton, VT, 2000.

Henderson, Bruce, *Fatal North: Murder and Survival on the First North Pole Expedition*, BruceHendersonBooks, Menlo Park, CA, 2014.

Henson, Matthew, *A Journey for the Ages: Matthew Henson and Robert Peary's Historic North Pole Expedition*, Skyhorse, New York, 2016 (first published 1912).

Herodotus, *The Histories*, trans. Aubrey de Sélincourt, Penguin Classics, London, 2003.

Heyerdahl, Thor, *The Kon-Tiki Expedition*, Simon & Schuster, New York, 1990.

Holland, Clive, *Arctic Exploration and Development c. 500 BC to 1915*, Frammuseet, Oslo, 2013 (first published 1994).

Huntford, Roland, *Nansen: The Explorer as Hero*, Gerald Duckworth, London, 1997.

Ingstad, Helge, *Pelsjegerliv blandt Nord-Kanadas indianere*, Gyldendal, Oslo, 1937.

Johansen, Hjalmar, *With Nansen in the North: A Record of the* Fram *Expedition in 1893–96*, Cambridge University Press, Cambridge, 2011.

Jølle, Harald Dag, *Nansen: Oppdageren. Vol. 1*, Gyldendal, Oslo, 2011.

Jorikson, Håkan, *S. A. Andrée ingenjören och polarfararen*, Carlsson, Stockholm, 2022.

Müller, Reidar, *Ild og is: En kort innføring i klimaets historie*, H. Aschehoug, Oslo, 2022.

Nansen, Fridtjof, *Eventyrlyst*, ed. Erling Kagge, Kagge Forlag, Oslo, 2011.

———, *Farthest North: The Incredible Three-Year Voyage to the Frozen Latitudes of the North*, Random House, New York, 1999.

———, *Nord i tåkeheimen: Utforskningen av jordens nordlige strøk i tidlige tider*, Jacob Dybwads, Kristiania, 1911.

Peary, Robert, *The North Pole*, Frederick A. Stokes, New York, 1910.

Pitzer, Andrea, *Icebound*, Simon & Schuster, London, 2021.

Ransmayr, Christoph, *The Terrors of Ice and Darkness*, Grove Press, New York, 1991.

Riffenburgh, Beau, *The Myth of the Explorer: The Press, Sensationalism, and Geographical Discovery*, Oxford University Press, Oxford, 1994.

Sides, Hampton, *In the Kingdom of Ice: The Grand and Terrible Polar Voyage of the USS* Jeannette, Doubleday, New York, 2014.

———, *The Wide, Wide Sea*, Doubleday, New York, 2024.

Sollinger, Günther, *S. A. Andrée: The Beginning of Polar Aviation 1895–1897*, Russian Academy of Sciences, Moscow, 2005.

Taylor, Andrew, *The World of Gerard Mercator: The Mapmaker Who Revolutionized Geography*, Walker Books, London, 2004.

Uusma, Bea, *The Expedition: A Love Story: Solving the Mystery of a Polar Tragedy*, Head of Zeus, London, 2014.

Vogt, Carl Emil, *Fridtjof Nansen: Mannen og verden*, J. W. Cappelens, Oslo, 2012.

Wisting, Alexander, *Roald Amundsen: Det største eventyret*, Kagge Forlag, Oslo, 2011.

Wulf, Andrea, *The Invention of Nature: The Adventures of Alexander von Humboldt, the Lost Hero of Science*, John Murray, London, 2015.

Zapffe, Peter Wessel, *On the Tragic*, trans. Ryan L. Showler, Peter Lang, Lausanne, 2024.

PERMISSIONS

pp. iv–v Anthony Fiala, photographer. *82 °N. Lat.* Arctic Ocean. March 1905. https://www.loc.gov/item/2007663116/.

p. vii Arctic Regions. 1908. Photograph. https://www.loc.gov/item/91729861/.

p. viii Nils Strindberg. 1897. Grenna Museum – Andréexpeditionen Polar-center. Photograph. https://digitaltmuseum.se/021016180085/anna-charlier-nils-strindbergs-fastmo-aterfunnet-pa-viton-1930.

p. xii Courtesy of Børge Ousland.

p. 9 Pictures Now / Alamy Stock Photo / NTB.

p. 54 F. Davignon, artist. Lewis & Brown, lithographer. Canada. 1840. Photograph. https://www.loc.gov/item/2003670428/.

p. 87 Courtesy of Børge Ousland.

p. 102 William A. Brady and Clay Meredith Greene. Strobridge & Co., lithographer. Cincinnati; New York. 1896. Photograph. https://www.loc.gov/item/2014637159/.

p. 140 Joseph Ferdinand Keppler, artist. *Science, or Sport?: Modern Spectacle After an Old Model* / J. Keppler. Europe. 1882. Published by Keppler & Schwarzmann. Photograph. https://www.loc.gov/item/2012647208/.

p. 168 Courtesy of Børge Ousland.

p. 187 Billedsentralen/NTB.

p. 209 Courtesy of Børge Ousland.

p. 239 Austin Mecklem, muralist. Carol M. Highsmith, photographer. Recorder of Deeds building, Washington, DC. 2010. Photograph. https://www.loc.gov/item/2010641718/.

p. 258 C. 1926. Photograph. https://www.loc.gov/item/2002721920/.

p. 279 C. June 17, 1926. Photograph. https://www.loc.gov/item/2002721911/.

p. 280 Spitsbergen Island, Norway. C. 1926. Photograph. https://www.loc.gov/item/2002721924/.

p. 281 C. July 14, 1926. Photograph. https://www.loc.gov/item/2002721901/.

p. 304 Courtesy of Børge Ousland.

p. 311 Courtesy of Børge Ousland.

p. 322 North Pole. 1905. Photograph. https://www.loc.gov/item/2012650303/.

p. 341 Spitsbergen Island, Norway. 1896 (or 1897). Photograph. https://www.loc.gov/item/2001700731/.

ABOUT THE AUTHOR

Norwegian explorer, philosopher, and writer Erling Kagge is the first person to complete the "Three Poles Challenge," reaching the North Pole, the South Pole, and the summit of Mount Everest on foot. Following this record-breaking feat, Kagge, already a lawyer, attended Cambridge University to study philosophy.

In recent years, Kagge has continued his adventures, both in the Arctic and in a series of urban explorations, including crossing New York City via its sewage, train, water, and subway tunnels.

Kagge is the author of such beloved books as *Silence: In the Age of Noise* and *Walking: One Step at a Time*. He has written for myriad publications, including the *Financial Times*, the *New York Times*, and the *Guardian*. In 1996, Kagge founded Kagge Forlag, which has become one of Norway's largest book publishers. He is also a dedicated art collector, having exhibited his collection of contemporary art in museums internationally to critical acclaim.

The father of three daughters, he lives in Oslo, Norway.

ABOUT THE TRANSLATOR

Kari Dickson is an award-winning literary translator from Norwegian. Her work includes crime fiction, literary fiction, children's books, theater, and nonfiction. She is also an occasional tutor in Norwegian language, literature, and translation at the University of Edinburgh, and has worked with the British Centre for Literary Translation, the National Centre for Writing, and the Translators Association.

NORTH
ATLANTIC OCEAN
GREENLAND
BAFFIN SEA
DAVIS STRAIT
HUDSON STRAIT
LABRADOR
ICELAND
FAROE ISLANDS
SPITZBERGEN
BRITISH ISLES
SCOTLAND
IRELAND
ENGLAND
Irish Channel
English Channel
NORTH SEA
HOLLAND
BELGIUM
HANOVER
PRUSSIA
BALTIC SEA
GULF OF BOTHNIA
GULF OF FINLAND
NORWAY
SWEDEN
LAPLAND
FINLAND
WHITE SEA
KARA SEA
NOVAYA ZEMLYA
Gulf of Obi
RUSSIA
EUROPE
WEST SIBERIA
Shetlands
Orkney
Jan Mayen
North Cape
Meridian 0° of Greenwich
Published at the Hydrographic Office of the Admiralty December 24th 1855
Sold by J. D. Potter Agent for the Admiralty Charts 31 Poultry & 11 King Street Tower Hill.
Engraved by J. & C. Walker.